Math

KS3-Year 7

Multiplication And Division
Addition, Subtraction

William. Education

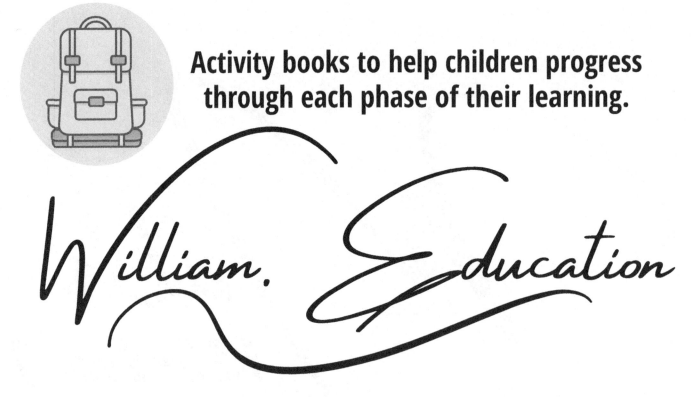

Activity books to help children progress through each phase of their learning.

Don't hesitate to give your opinion (constructive ;-)) and your ideas for improvement after your purchase, because I really want to offer quality, Have fun.

Peaperback ASIN : 9798394040122

TABLE OF CONTENTS

 **Addition** 20 worksheets
30 problems per sheet

Subtraction 20 worksheets
30 problems per sheet

Multiplication 20 worksheets
30 problems per sheet

Division 20 worksheets
40 problems per sheet

Solutions to the exercises can be found in the back of the book

Addition ✚

20 worksheets

30 problems per sheet

William. Education

Examples

Example 1 (2-digit numbers)

1
```
   98
+  63
―――――
```

2
```
    1
   98
+  63
―――――
    1
```

3
```
    1
   98
+  63
―――――
  161
```

Example 2 (3-digit numbers)

1
```
   286
+   82
――――――
```

2
```
   286
+   82
――――――
     8
```

3
```
     1
   286
+   82
――――――
    68
```

4
```
    1
   286
+   82
――――――
   368
```

Example 3 (3-digit numbers)

1
```
   387
+  688
――――――
```

2
```
     1
   387
+  688
――――――
     5
```

3
```
    1 1
   387
+  688
――――――
    75
```

4
```
    1 1
   387
+  688
――――――
  1075
```

Example 4 (2-digit numbers)

$+\overset{1}{}$ 87 88 **175**	$+\overset{1}{}$ 95 38 **133**	+ 24 25 **49**	$+\overset{1}{}$ 81 49 **130**	+ 82 74 **156**
$+\overset{1}{}$ 32 19 **51**	+ 83 36 **119**	+ 50 40 **90**	$+\overset{1}{}$ 16 18 **34**	$+\overset{1}{}$ 48 23 **71**

Example 5 (3-digit numbers)

$+\overset{11}{}$ 487 778 **1265**	$+\overset{11}{}$ 557 868 **1425**	+ 721 975 **1696**	$+\overset{11}{}$ 989 54 **1043**	$+\overset{11}{}$ 453 49 **502**
$+\overset{11}{}$ 675 569 **1244**	$+\overset{1}{}$ 454 929 **1383**	+ 621 747 **1368**	+ 101 661 **762**	$+\overset{11}{}$ 538 783 **1321**

Example 6 (4-digit numbers)

$+\overset{1}{}$ 2168 1080 **3248**	$+\overset{11}{}$ 4876 8143 **13019**	$+\overset{1}{}$ 7284 7323 **14607**	+ 3753 244 **3997**	$+\overset{111}{}$ 4449 855 **5304**
$+\overset{11}{}$ 5278 1572 **6850**	$+\overset{1}{}$ 7681 9247 **16928**	+ 4572 7323 **11895**	+ 7360 535 **7895**	$+\overset{1}{}$ 8270 384 **8654**

Addition 3-Digit Numbers

Day: **Time:**   **Score:**   **/30**

1). 895
    +675
    _____

2). 898
    +113
    _____

3). 391
    +262
    _____

4). 705
    +303
    _____

5). 470
    +879
    _____

6). 449
    +160
    _____

7). 752
    +787
    _____

8). 490
    +862
    _____

9). 586
    +905
    _____

10). 875
    +294
    _____

11). 142
    +832
    _____

12). 999
    +605
    _____

13). 386
    +276
    _____

14). 277
    +948
    _____

15). 772
    +440
    _____

16). 571
    +400
    _____

17). 218
    +841
    _____

18). 942
    +349
    _____

19). 835
    +407
    _____

20). 397
    +782
    _____

21). 224
    +697
    _____

22). 977
    +269
    _____

23). 882
    +744
    _____

24). 945
    +966
    _____

25). 634
    +529
    _____

26). 632
    +931
    _____

27). 711
    +846
    _____

28). 690
    +505
    _____

29). 108
    +332
    _____

30). 153
    +628
    _____

1

Addition 3-Digit Numbers

Day: **Time:** **Score:** **/30**

1). 956
+793
_____

2). 127
+629
_____

3). 182
+390
_____

4). 640
+612
_____

5). 849
+158
_____

6). 983
+913
_____

7). 586
+836
_____

8). 720
+691
_____

9). 601
+426
_____

10). 356
+570
_____

11). 424
+645
_____

12). 349
+387
_____

13). 704
+199
_____

14). 791
+411
_____

15). 509
+249
_____

16). 258
+874
_____

17). 274
+975
_____

18). 674
+395
_____

19). 131
+129
_____

20). 315
+144
_____

21). 699
+599
_____

22). 355
+496
_____

23). 842
+583
_____

24). 316
+615
_____

25). 448
+287
_____

26). 819
+547
_____

27). 872
+688
_____

28). 793
+592
_____

29). 950
+831
_____

30). 228
+157
_____

Addition 3-Digit Numbers

1).   762
+696
_____

2).   157
+960
_____

3).   193
+660
_____

4).   386
+956
_____

5).   338
+211
_____

6).   312
+513
_____

7).   667
+947
_____

8).   753
+657
_____

9).   536
+785
_____

10).   636
+378
_____

11).   796
+719
_____

12).   499
+966
_____

13).   183
+405
_____

14).   563
+784
_____

15).   832
+776
_____

16).   741
+495
_____

17).   588
+130
_____

18).   385
+673
_____

19).   688
+735
_____

20).   853
+435
_____

21).   307
+234
_____

22).   137
+781
_____

23).   605
+509
_____

24).   518
+539
_____

25).   372
+282
_____

26).   608
+134
_____

27).   520
+483
_____

28).   368
+724
_____

29).   781
+143
_____

30).   152
+771
_____

Addition 3-Digit Numbers

1).   315
+212
_____

2).   641
+111
_____

3).   499
+770
_____

4).   783
+264
_____

5).   492
+466
_____

6).
531
+591
_____

7).
180
+619
_____

8).
539
+327
_____

9).
903
+954
_____

10).
382
+491
_____

11).
997
+327
_____

12).
276
+548
_____

13).
886
+449
_____

14).
413
+196
_____

15).
318
+623
_____

16).
697
+748
_____

17).
727
+542
_____

18).
213
+435
_____

19).
380
+977
_____

20).
128
+274
_____

21).
608
+346
_____

22).
682
+874
_____

23).
221
+954
_____

24).
143
+776
_____

25).
507
+816
_____

26).
740
+199
_____

27).
472
+372
_____

28).
922
+918
_____

29).
415
+966
_____

30).
155
+482
_____

Addition 3-Digit Numbers

Day: **Time:**   **Score:**   /30

1). 801
    +428
    _____

2). 435
    +603
    _____

3). 855
    +288
    _____

4). 466
    +491
    _____

5). 359
    +541
    _____

6). 444
    +621
    _____

7). 704
    +982
    _____

8). 404
    +861
    _____

9). 907
    +356
    _____

10). 457
     +315
     _____

11). 433
     +315
     _____

12). 780
     +939
     _____

13). 841
     +541
     _____

14). 661
     +914
     _____

15). 879
     +125
     _____

16). 470
     +119
     _____

17). 637
     +948
     _____

18). 105
     +422
     _____

19). 799
     +464
     _____

20). 121
     +315
     _____

21). 266
     +753
     _____

22). 437
     +781
     _____

23). 784
     +412
     _____

24). 331
     +575
     _____

25). 179
     +823
     _____

26). 992
     +121
     _____

27). 578
     +104
     _____

28). 746
     +216
     _____

29). 214
     +708
     _____

30). 134
     +565
     _____

Addition 4-Digit Numbers

1). 3446
    +6220
    _____

2). 8688
    +7389
    _____

3). 6885
    +5994
    _____

4). 8673
    +8547
    _____

5). 3167
    +2456
    _____

6). 3795
    +8628
    _____

7). 5195
    +4131
    _____

8). 5878
    +7764
    _____

9). 1739
    +7235
    _____

10). 7508
    +8670
    _____

11). 3207
    +9213
    _____

12). 4019
    +3539
    _____

13). 2863
    +7189
    _____

14). 1269
    +3416
    _____

15). 6941
    +5869
    _____

16). 6910
    +4211
    _____

17). 7453
    +2101
    _____

18). 8577
    +3639
    _____

19). 1849
    +8123
    _____

20). 3581
    +2142
    _____

21). 3897
    +8495
    _____

22). 9232
    +4301
    _____

23). 7263
    +9881
    _____

24). 1887
    +7307
    _____

25). 3347
    +3109
    _____

26). 5168
    +8842
    _____

27). 3828
    +9397
    _____

28). 2583
    +8097
    _____

29). 8262
    +5264
    _____

30). 2936
    +8733
    _____

Addition 4-Digit Numbers

1). 9903
  +6755
  _____

2). 9886
  +7395
  _____

3). 4455
  +1547
  _____

4). 1877
  +1341
  _____

5). 5800
  +2484
  _____

6). 3912
  +3186
  _____

7). 1522
  +3967
  _____

8). 1116
  +8832
  _____

9). 8041
  +5832
  _____

10). 5510
  +4671
  _____

11). 6495
  +2409
  _____

12). 9376
  +7348
  _____

13). 9251
  +8925
  _____

14). 2113
  +7012
  _____

15). 9815
  +7208
  _____

16). 4321
  +1694
  _____

17). 3860
  +4497
  _____

18). 3375
  +1926
  _____

19). 3400
  +3599
  _____

20). 7164
  +7407
  _____

21). 7883
  +1787
  _____

22). 5981
  +7321
  _____

23). 2143
  +8579
  _____

24). 3762
  +5936
  _____

25). 5668
  +2996
  _____

26). 9035
  +6201
  _____

27). 9624
  +8975
  _____

28). 6934
  +3384
  _____

29). 9435
  +9375
  _____

30). 7612
  +6247
  _____

Addition 4-Digit Numbers

1). 4708
    +5477
    _____

2). 3636
    +4897
    _____

3). 5552
    +8274
    _____

4). 7295
    +4354
    _____

5). 7353
    +6041
    _____

6). 1015
    +7593
    _____

7). 3264
    +1791
    _____

8). 8849
    +3584
    _____

9). 8957
    +7041
    _____

10). 6575
    +2292
    _____

11). 8744
    +2874
    _____

12). 5173
    +4054
    _____

13). 6430
    +3636
    _____

14). 2202
    +4407
    _____

15). 7671
    +2940
    _____

16). 2259
    +5386
    _____

17). 5252
    +5183
    _____

18). 8516
    +2153
    _____

19). 7884
    +1216
    _____

20). 1272
    +3122
    _____

21). 9078
    +2177
    _____

22). 5563
    +7382
    _____

23). 3178
    +2717
    _____

24). 3136
    +9337
    _____

25). 9860
    +3049
    _____

26). 5522
    +4955
    _____

27). 7299
    +6286
    _____

28). 4492
    +5909
    _____

29). 3681
    +6533
    _____

30). 3948
    +2207
    _____

Addition 4-Digit Numbers

1).   2285
    +9209
____

2).   2680
    +3603
____

3).   7840
    +8669
____

4).   7061
    +7755
____

5).   4161
    +9019
____

6).
    2198
    +5765
____

7).
    1614
    +7741
____

8).
    4023
    +6362
____

9).
    6458
    +4174
____

10).
    1476
    +6239
____

11).
    4449
    +2441
____

12).
    9896
    +5838
____

13).
    1178
    +1455
____

14).
    2385
    +9133
____

15).
    8176
    +7952
____

16).
    2436
    +1990
____

17).
    7516
    +6986
____

18).
    2838
    +4329
____

19).
    7607
    +3394
____

20).
    3247
    +1925
____

21).
    3930
    +5083
____

22).
    9835
    +9095
____

23).
    6705
    +9813
____

24).
    7107
    +1955
____

25).
    5859
    +6347
____

26).
    1343
    +9185
____

27).
    9214
    +5477
____

28).
    4078
    +3441
____

29).
    5513
    +5414
____

30).
    1722
    +2653
____

Addition 4-Digit Numbers

  Time: Score: /30

1). 4477
    +6975
    _____

2). 5337
    +8154
    _____

3). 9260
    +7449
    _____

4). 8999
    +7992
    _____

5). 1260
    +8817
    _____

6). 3043
    +9725
    _____

7). 8281
    +5777
    _____

8). 9751
    +1053
    _____

9). 8330
    +7679
    _____

10). 7306
    +7892
    _____

11). 5503
    +8209
    _____

12). 5704
    +2141
    _____

13). 9086
    +6818
    _____

14). 9657
    +2500
    _____

15). 6952
    +3821
    _____

16). 6250
    +5617
    _____

17). 3259
    +9023
    _____

18). 5808
    +9732
    _____

19). 4211
    +4704
    _____

20). 2540
    +4427
    _____

21). 6365
    +1068
    _____

22). 2047
    +9305
    _____

23). 2533
    +5564
    _____

24). 4722
    +7172
    _____

25). 4862
    +4023
    _____

26). 9381
    +4812
    _____

27). 3071
    +3226
    _____

28). 8223
    +8530
    _____

29). 6136
    +9326
    _____

30). 5351
    +6433
    _____

Addition 5-Digit Numbers

1). 23461
    +4192
    _____

2). 36127
    +4929
    _____

3). 66806
    +1414
    _____

4). 71796
    +5239
    _____

5). 37312
    +1582
    _____

6).
    52827
    +7996
    _____

7).
    17488
    +5180
    _____

8).
    95313
    +6204
    _____

9).
    76299
    +5257
    _____

10).
    62132
    +7090
    _____

11).
    75307
    +2393
    _____

12).
    71011
    +7610
    _____

13).
    59505
    +4754
    _____

14).
    68927
    +6471
    _____

15).
    70951
    +2794
    _____

16).
    24508
    +7058
    _____

17).
    31446
    +7488
    _____

18).
    66588
    +4941
    _____

19).
    47808
    +6477
    _____

20).
    27558
    +5572
    _____

21).
    89512
    +6111
    _____

22).
    64733
    +2354
    _____

23).
    63887
    +1215
    _____

24).
    51640
    +3543
    _____

25).
    28467
    +8309
    _____

26).
    13769
    +7315
    _____

27).
    43555
    +6150
    _____

28).
    20162
    +7915
    _____

29).
    70653
    +9707
    _____

30).
    33002
    +6007
    _____

Addition 5-Digit Numbers

1). 34800
    +8679
    _____

2). 51621
    +2166
    _____

3). 95856
    +8311
    _____

4). 38522
    +4662
    _____

5). 45588
    +4010
    _____

6). 89668
    +7068
    _____

7). 67133
    +6763
    _____

8). 71697
    +8650
    _____

9). 13734
    +5719
    _____

10). 13367
     +3493
     _____

11). 81631
     +5815
     _____

12). 95800
     +8973
     _____

13). 63241
     +6414
     _____

14). 20021
     +2437
     _____

15). 80969
     +1031
     _____

16). 63143
     +8882
     _____

17). 83790
     +5223
     _____

18). 62106
     +6223
     _____

19). 44550
     +5503
     _____

20). 50509
     +6627
     _____

21). 85845
     +7447
     _____

22). 49708
     +4491
     _____

23). 36284
     +2237
     _____

24). 70124
     +9864
     _____

25). 96946
     +8543
     _____

26). 42495
     +4480
     _____

27). 99957
     +3998
     _____

28). 11651
     +4160
     _____

29). 17209
     +4509
     _____

30). 67340
     +9992
     _____

Addition 5-Digit Numbers

Day: **Time:** **Score:** /30

1). 16932
     +6494
     _____

2). 62795
     +1669
     _____

3). 12922
     +8281
     _____

4). 20313
     +9471
     _____

5). 49189
     +8720
     _____

6). 20980
     +3673
     _____

7). 55147
     +7753
     _____

8). 71612
     +4835
     _____

9). 50225
     +4343
     _____

10). 48366
     +5576
     _____

11). 53265
     +1736
     _____

12). 41381
     +4056
     _____

13). 12560
     +6271
     _____

14). 73617
     +2912
     _____

15). 60550
     +1205
     _____

16). 44083
     +4406
     _____

17). 12443
     +2951
     _____

18). 42985
     +2858
     _____

19). 89919
     +7052
     _____

20). 25167
     +1116
     _____

21). 29445
     +2966
     _____

22). 78698
     +5964
     _____

23). 19713
     +2447
     _____

24). 87424
     +3850
     _____

25). 94264
     +4889
     _____

26). 89898
     +2281
     _____

27). 63415
     +5542
     _____

28). 92192
     +1543
     _____

29). 94312
     +7861
     _____

30). 38839
     +1145
     _____

Addition 5-Digit Numbers

1). 32512
   +9503
   _____

2). 86155
   +5748
   _____

3). 78396
   +4437
   _____

4). 52826
   +1722
   _____

5). 47992
   +6540
   _____

6). 92449
   +9670
   _____

7). 92017
   +2738
   _____

8). 97303
   +7352
   _____

9). 49987
   +4865
   _____

10). 46018
    +5093
    _____

11). 91044
    +4815
    _____

12). 75802
    +3243
    _____

13). 96717
    +4482
    _____

14). 23909
    +8347
    _____

15). 86542
    +2828
    _____

16). 14923
    +9937
    _____

17). 72908
    +7040
    _____

18). 10383
    +4059
    _____

19). 18496
    +5759
    _____

20). 83863
    +6744
    _____

21). 58224
    +2794
    _____

22). 47633
    +7618
    _____

23). 87787
    +2548
    _____

24). 76784
    +6796
    _____

25). 82506
    +1654
    _____

26). 12820
    +8985
    _____

27). 83505
    +7935
    _____

28). 11596
    +8465
    _____

29). 43024
    +3868
    _____

30). 49745
    +2236
    _____

Addition 5-Digit Numbers

1). 43714
+4044

2). 74258
+7356

3). 85507
+1363

4). 14734
+6027

5). 60585
+5086

6). 73816
+7922

7). 67255
+8558

8). 95941
+4170

9). 97318
+7380

10). 81540
+9381

11). 85783
+4932

12). 43491
+6849

13). 98780
+9057

14). 78741
+5302

15). 57859
+3442

16). 11689
+8027

17). 23563
+2643

18). 92603
+6915

19). 32330
+8025

20). 52605
+4135

21). 88613
+8108

22). 66555
+9705

23). 38419
+3948

24). 96581
+9150

25). 11397
+9188

26). 51892
+5847

27). 56046
+3411

28). 65039
+3393

29). 66702
+5217

30). 10875
+5528

15

Addition 5-Digit Numbers

Day: Time: Score: /30

1). 60736
+42215
_____

2). 98191
+73003
_____

3). 23264
+46305
_____

4). 47241
+24027
_____

5). 12068
+94257
_____

6).
34987
+25392
_____

7).
59114
+36273
_____

8).
91574
+76735
_____

9).
46094
+35984
_____

10).
96867
+17399
_____

11).
10734
+98382
_____

12).
39160
+53131
_____

13).
14946
+66931
_____

14).
99075
+49613
_____

15).
74343
+57116
_____

16).
46553
+86989
_____

17).
23975
+11621
_____

18).
27500
+52001
_____

19).
69548
+91827
_____

20).
39360
+93390
_____

21).
16920
+71393
_____

22).
81648
+21820
_____

23).
35165
+80302
_____

24).
18304
+37731
_____

25).
20151
+41947
_____

26).
23624
+59245
_____

27).
56231
+31590
_____

28).
12487
+17556
_____

29).
63369
+21816
_____

30).
28604
+69042
_____

Addition 5-Digit Numbers

1). 63751
+98022

2). 92546
+15647

3). 46196
+81825

4). 26659
+78236

5). 66900
+21054

6). 26698
+90398

7). 53803
+44782

8). 69900
+47213

9). 70351
+28532

10). 57170
+95235

11). 42998
+65149

12). 35294
+96722

13). 80408
+79652

14). 86064
+44415

15). 72935
+20983

16). 84937
+27313

17). 67595
+72757

18). 61774
+41623

19). 19422
+35874

20). 10161
+41305

21). 84808
+68163

22). 79804
+39520

23). 26262
+70109

24). 48608
+65672

25). 60863
+91155

26). 77163
+82718

27). 36607
+28055

28). 38573
+35943

29). 41836
+65360

30). 10847
+37503

Day: Time: Score: /30

1).   64400
   +69715
_____

2). 58929
  +26221
_____

3). 24821
  +33344
_____

4).   15643
   +78607
_____

5). 39376
  +89429
_____

6).   67409
  +61171
_____

7).   82251
  +69158
_____

8).   72727
  +24609
_____

9).   81892
  +89721
_____

10). 36551
  +38430
_____

11). 14606
  +31156
_____

12). 73245
  +69579
_____

13). 55327
  +49719
_____

14). 95537
  +72834
_____

15). 31558
  +46518
_____

16). 37702
  +86475
_____

17). 20083
  +89380
_____

18). 39300
  +61724
_____

19). 31154
  +49290
_____

20). 20452
  +48592
_____

21). 21568
  +10085
_____

22). 10883
  +63944
_____

23). 86274
  +21598
_____

24). 62695
  +69816
_____

25). 87997
  +11903
_____

26). 41235
  +31325
_____

27). 85704
  +53957
_____

28). 99283
  +73878
_____

29). 65719
  +35518
_____

30). 37286
  +58798
_____

Addition 5-Digit Numbers

Day: **Time:**   **Score:**   /30

1).   22286
    +49698
_____

2). 84630
    +22536
_____

3). 93851
    +75404
_____

4).   21706
    +34738
_____

5). 81772
    +23773
_____

6).
   44800
  +19930
_____

7).
   66482
  +27263
_____

8).
   73947
  +96029
_____

9).
   45138
  +41157
_____

10).
   33103
  +20311
_____

11).
   41205
  +52179
_____

12).
   75963
  +24056
_____

13).
   69193
  +30653
_____

14).
   57652
  +49831
_____

15).
   55506
  +55673
_____

16).
   26519
  +20506
_____

17).
   89552
  +79242
_____

18).
   33055
  +56324
_____

19).
   24152
  +95848
_____

20).
   52317
  +42639
_____

21).
   40599
  +98508
_____

22).
   22121
  +26315
_____

23).
   81691
  +80976
_____

24).
   51755
  +13189
_____

25).
   75568
  +68503
_____

26).
   45001
  +22477
_____

27).
   44217
  +76315
_____

28).
   86835
  +86593
_____

29).
   43174
  +95782
_____

30).
   45333
  +36404
_____

Addition 5-Digit Numbers

1). 40557
 +34775
_____

2). 68843
 +36168
_____

3). 91599
 +60301
_____

4). 44427
 +41318
_____

5). 32786
 +34950
_____

6).
 74938
+71500
_____

7).
 48552
+55310
_____

8).
 86089
+74809
_____

9).
 75758
+58729
_____

10).
 68165
+75123
_____

11).
 69139
+80148
_____

12).
 83853
+96254
_____

13).
 98179
+53185
_____

14).
 79029
+47173
_____

15).
 14695
+21743
_____

16).
 39363
+48773
_____

17).
 57926
+27812
_____

18).
 60085
+10340
_____

19).
 27929
+99657
_____

20).
 82328
+15678
_____

21).
 48704
+92104
_____

22).
 37092
+84653
_____

23).
 29592
+47253
_____

24).
 47057
+25823
_____

25).
 37253
+19647
_____

26).
 47231
+99768
_____

27).
 20784
+64619
_____

28).
 47727
+80764
_____

29).
 79395
+22364
_____

30).
 65846
+96509
_____

Subtraction

20 worksheets
30 problems per sheet

William. Education

Examples

Example 1 (2-digit numbers)

1
```
   95
-
   46
```

2
```
   8 15
   9̷5̷
     ↓
   46
    9
```

3
```
   8 15
   9̷5̷
     ↓
   46
   49
```

Example 2 (3-digit numbers)

1
```
   813
-
    75
```

2
```
   0 13
   8̷1̷3
      ↓
    75
     8
```

3
```
   7 10 13
   8̷1̷3̷
       ↓
    75
    38
```

4
```
   7 10 13
   8̷1̷3̷
       ↓
    75
   738
```

Example 3 (3-digit numbers)

1
```
   600
-
   464
```

2
```
      9
   5 10 10
   6̷00
      ↓
   464
     6
```

3
```
      9
   5 10 10
   6̷0̷0
      ↓
   464
    36
```

4
```
      9
   5 10 10
   6̷0̷0̷
      ↓
   464
   136
```

Example 4 (2-digit numbers)

	8 15			7 14					3 10			
-	9̶5		-	8̶4		-	87		-	4̶0		-	56
	66			18			70			22			26
	29			**66**			**17**			**18**			**30**

			4 16					
-	88	-	5̶6	-	29	-	99	-	59
	28		18		20		44		51
	60		**38**		**9**		**55**		**8**

Example 5 (3-digit numbers)

	4 10		1 11		5 14		6 11		9 7 10 15
-	35̶0	-	2̶16	-	26̶4	-	7̶19	-	8̶0̶5
	236		193		229		64		67
	114		**23**		**35**		**655**		**738**

	8 ¹² 2 10		4 12		4 13		4 10		13 4 3 14
-	9̶3̶0	-	7̶5̶2	-	9̶5̶3	-	6̶5̶0	-	5̶4̶4
	786		126		138		44		58
	144		**626**		**815**		**606**		**486**

Example 6 (4-digit numbers)

	3 10						4 13		13 3 3 14
-	1̶404	-	3578	-	5678	-	37̶5̶3	-	4̶4̶4̶9
	1080		1563		2528		244		855
	324		**2015**		**3150**		**3509**		**3594**

	8 11		1 14		4 10		6 13 5 10		11 16 7 1 6 10
-	1̶9̶19	-	1̶243	-	5̶055	-	7̶3̶6̶0	-	8̶2̶7̶0
	1632		1191		1211		535		384
	287		**52**		**3844**		**6825**		**7886**

Subtraction 3-Digit Numbers

Day: **Time:**   **Score:**   /30

1).   127
     -107
_____

2).   259
     -158
_____

3).   838
     -406
_____

4).   611
     -364
_____

5).   595
     -434
_____

6).   488
     -260
_____

7).   718
     -204
_____

8).   742
     -351
_____

9).   231
     -207
_____

10).   770
     -659
_____

11).   487
     -374
_____

12).   401
     -194
_____

13).   481
     -239
_____

14).   380
     -234
_____

15).   868
     -824
_____

16).   972
     -590
_____

17).   838
     -194
_____

18).   861
     -739
_____

19).   589
     -571
_____

20).   181
     -180
_____

21).   712
     -565
_____

22).   484
     -480
_____

23).   201
     -109
_____

24).   465
     -217
_____

25).   227
     -128
_____

26).   261
     -251
_____

27).   434
     -164
_____

28).   565
     -155
_____

29).   457
     -383
_____

30).   251
     -173
_____

Subtraction 3-Digit Numbers

Day: _____ ⏱ Time: _____ Score: _____ /30

1). 966
    -586
    _____

2). 157
    -143
    _____

3). 535
    -172
    _____

4). 894
    -105
    _____

5). 623
    -257
    _____

6). 962
    -678
    _____

7). 902
    -663
    _____

8). 212
    -138
    _____

9). 947
    -637
    _____

10). 427
     -379
     _____

11). 688
     -337
     _____

12). 556
     -405
     _____

13). 945
     -869
     _____

14). 668
     -155
     _____

15). 818
     -581
     _____

16). 527
     -314
     _____

17). 486
     -480
     _____

18). 436
     -108
     _____

19). 390
     -318
     _____

20). 383
     -113
     _____

21). 315
     -312
     _____

22). 661
     -201
     _____

23). 318
     -166
     _____

24). 763
     -505
     _____

25). 725
     -364
     _____

26). 329
     -234
     _____

27). 287
     -148
     _____

28). 266
     -108
     _____

29). 181
     -130
     _____

30). 900
     -749
     _____

Subtraction 3-Digit Numbers

1). 158
    -133
    _____

2). 877
    -168
    _____

3). 153
    -121
    _____

4). 601
    -260
    _____

5). 807
    -155
    _____

6). 643
    -329
    _____

7). 310
    -138
    _____

8). 103
    -100
    _____

9). 272
    -235
    _____

10). 336
    -114
    _____

11). 654
    -555
    _____

12). 338
    -278
    _____

13). 236
    -143
    _____

14). 357
    -103
    _____

15). 448
    -437
    _____

16). 955
    -386
    _____

17). 723
    -214
    _____

18). 586
    -294
    _____

19). 853
    -322
    _____

20). 513
    -400
    _____

21). 244
    -215
    _____

22). 660
    -449
    _____

23). 788
    -695
    _____

24). 278
    -244
    _____

25). 123
    -114
    _____

26). 519
    -109
    _____

27). 798
    -464
    _____

28). 524
    -481
    _____

29). 675
    -410
    _____

30). 845
    -194
    _____

Subtraction 3-Digit Numbers

1).   196
   -127
_____

2).   188
   -124
_____

3).   520
   -101
_____

4).   836
   -659
_____

5).   361
   -181
_____

6).   157
   -107
_____

7).   575
   -485
_____

8).   109
   -102
_____

9).   903
   -799
_____

10).   367
   -105
_____

11).   104
   -100
_____

12).   650
   -591
_____

13).   763
   -675
_____

14).   636
   -229
_____

15).   873
   -638
_____

16).   212
   -196
_____

17).   849
   -636
_____

18).   215
   -155
_____

19).   400
   -280
_____

20).   666
   -623
_____

21).   685
   -191
_____

22).   660
   -509
_____

23).   661
   -126
_____

24).   753
   -654
_____

25).   287
   -128
_____

26).   367
   -318
_____

27).   142
   -135
_____

28).   855
   -710
_____

29).   707
   -350
_____

30).   170
   -116
_____

Subtraction 3-Digit Numbers

1).   464
    -232

2).   304
    -150

3).   264
    -235

4).   965
    -180

5).   612
    -602

6).   486
    -431

7).   582
    -568

8).   712
    -158

9).   348
    -172

10).   682
    -641

11).   126
    -121

12).   578
    -431

13).   360
    -313

14).   334
    -181

15).   183
    -166

16).   143
    -117

17).   969
    -795

18).   971
    -854

19).   236
    -113

20).   932
    -454

21).   931
    -702

22).   601
    -468

23).   273
    -220

24).   488
    -117

25).   155
    -107

26).   409
    -401

27).   748
    -138

28).   699
    -351

29).   854
    -587

30).   511
    -491

Subtraction 4-Digit Numbers

1). 5645
    -3637
    _____

2). 4241
    -1206
    _____

3). 6545
    -1756
    _____

4). 6816
    -3311
    _____

5). 7686
    -2436
    _____

6). 1053
    -1010
    _____

7). 9982
    -6451
    _____

8). 1480
    -1076
    _____

9). 9059
    -4556
    _____

10). 3466
    -3208
    _____

11). 9717
    -1103
    _____

12). 1239
    -1192
    _____

13). 9694
    -7353
    _____

14). 8432
    -6354
    _____

15). 1255
    -1181
    _____

16). 6488
    -3076
    _____

17). 3472
    -2533
    _____

18). 2755
    -1351
    _____

19). 4927
    -4676
    _____

20). 3431
    -2206
    _____

21). 5726
    -4387
    _____

22). 1868
    -1642
    _____

23). 1363
    -1097
    _____

24). 5885
    -5655
    _____

25). 8963
    -8077
    _____

26). 2873
    -2046
    _____

27). 9886
    -9388
    _____

28). 2449
    -1434
    _____

29). 6490
    -2046
    _____

30). 3363
    -3181
    _____

Subtraction 4-Digit Numbers

1). 2453
    -1333
_____

2). 9178
    -3233
_____

3). 3435
    -1938
_____

4). 5651
    -4603
_____

5). 4109
    -2988
_____

6). 4453
    -3131
_____

7). 2711
    -2344
_____

8). 6502
    -5871
_____

9). 6638
    -3765
_____

10). 3404
    -2296
_____

11). 8115
    -3380
_____

12). 6412
    -5765
_____

13). 8600
    -3325
_____

14). 6414
    -2760
_____

15). 1245
    -1225
_____

16). 9354
    -1535
_____

17). 4901
    -3800
_____

18). 8298
    -8015
_____

19). 7912
    -2138
_____

20). 9333
    -1106
_____

21). 6376
    -5478
_____

22). 6644
    -5411
_____

23). 3664
    -1963
_____

24). 2395
    -1683
_____

25). 9695
    -8522
_____

26). 6493
    -5802
_____

27). 1912
    -1320
_____

28). 5912
    -3326
_____

29). 6934
    -2387
_____

30). 1037
    -1028
_____

Subtraction 4-Digit Numbers

Day: Time: Score: /30

1). 7528
    -6972
    ‾‾‾‾‾

2). 5143
    -2907
    ‾‾‾‾‾

3). 7012
    -4053
    ‾‾‾‾‾

4). 6453
    -3323
    ‾‾‾‾‾

5). 3344
    -3120
    ‾‾‾‾‾

6). 7578
    -7029
    ‾‾‾‾‾

7). 2797
    -1399
    ‾‾‾‾‾

8). 8523
    -2011
    ‾‾‾‾‾

9). 7004
    -6528
    ‾‾‾‾‾

10). 3867
     -1077
     ‾‾‾‾‾

11). 5046
     -1057
     ‾‾‾‾‾

12). 5999
     -3344
     ‾‾‾‾‾

13). 6277
     -2565
     ‾‾‾‾‾

14). 1639
     -1281
     ‾‾‾‾‾

15). 3811
     -2912
     ‾‾‾‾‾

16). 9151
     -4257
     ‾‾‾‾‾

17). 6992
     -6525
     ‾‾‾‾‾

18). 8705
     -7218
     ‾‾‾‾‾

19). 9092
     -5826
     ‾‾‾‾‾

20). 9757
     -4570
     ‾‾‾‾‾

21). 7714
     -2083
     ‾‾‾‾‾

22). 1720
     -1333
     ‾‾‾‾‾

23). 7062
     -3196
     ‾‾‾‾‾

24). 1885
     -1517
     ‾‾‾‾‾

25). 5340
     -1072
     ‾‾‾‾‾

26). 6873
     -2141
     ‾‾‾‾‾

27). 1009
     -1003
     ‾‾‾‾‾

28). 7928
     -7642
     ‾‾‾‾‾

29). 2533
     -1515
     ‾‾‾‾‾

30). 5350
     -3964
     ‾‾‾‾‾

Subtraction 4-Digit Numbers

1). 2367
-2167
___

2). 6792
-5980
___

3). 1683
-1591
___

4). 6526
-6303
___

5). 5366
-3036
___

6). 3455
-1291
___

7). 3528
-3374
___

8). 5544
-5284
___

9). 4733
-3287
___

10). 7137
-1168
___

11). 1773
-1549
___

12). 4213
-2176
___

13). 5161
-5066
___

14). 5287
-3256
___

15). 4653
-2496
___

16). 3050
-1564
___

17). 6110
-2461
___

18). 4225
-1541
___

19). 1027
-1020
___

20). 9163
-1897
___

21). 2312
-1441
___

22). 8347
-1516
___

23). 6284
-3258
___

24). 8354
-7692
___

25). 6412
-4430
___

26). 3800
-3174
___

27). 7982
-3886
___

28). 7444
-5036
___

29). 4783
-3012
___

30). 8957
-2747
___

Subtraction 4-Digit Numbers

1).   8786
      -3052

2).   3677
      -3475

3).   4537
      -3260

4).   4334
      -1948

5).   6790
      -2915

6).   3709
      -1880

7).   9155
      -7233

8).   9830
      -3321

9).   6641
      -4843

10).   4133
      -3217

11).   5321
      -2810

12).   6192
      -4523

13).   7977
      -2002

14).   8648
      -8035

15).   4239
      -4044

16).   7721
      -3598

17).   3658
      -2353

18).   7541
      -1527

19).   6164
      -2278

20).   1043
      -1016

21).   1579
      -1369

22).   4288
      -3122

23).   7208
      -2367

24).   1935
      -1095

25).   6037
      -5463

26).   3996
      -1175

27).   4739
      -4261

28).   4183
      -4025

29).   3531
      -2608

30).   6891
      -1468

Subtraction 5-Digit Numbers

1). 19892
    -7336
    _____

2). 65513
    -2009
    _____

3). 75270
    -2245
    _____

4). 75726
    -9877
    _____

5). 85023
    -1127
    _____

6). 87911
    -5920
    _____

7). 65590
    -9226
    _____

8). 48254
    -2275
    _____

9). 34159
    -3969
    _____

10). 76289
    -9880
    _____

11). 79704
    -7338
    _____

12). 62073
    -4124
    _____

13). 31624
    -8190
    _____

14). 60047
    -7247
    _____

15). 69242
    -1499
    _____

16). 96009
    -4766
    _____

17). 15869
    -6562
    _____

18). 58585
    -6888
    _____

19). 93156
    -8795
    _____

20). 98127
    -4483
    _____

21). 44998
    -6605
    _____

22). 34039
    -3675
    _____

23). 85384
    -7205
    _____

24). 56225
    -7556
    _____

25). 71308
    -9958
    _____

26). 97537
    -7792
    _____

27). 10051
    -4989
    _____

28). 11922
    -6173
    _____

29). 10018
    -5340
    _____

30). 17269
    -4236
    _____

Subtraction 5-Digit Numbers

1). 58601
     -4927
_____

2). 24962
     -5302
_____

3). 16769
     -3577
_____

4). 51897
     -5523
_____

5). 75123
     -4082
_____

6). 93675
     -1041
_____

7). 28436
     -3394
_____

8). 49671
     -5927
_____

9). 65876
     -6945
_____

10). 72547
     -9494
_____

11). 46359
     -4947
_____

12). 16760
     -2243
_____

13). 97265
     -5037
_____

14). 49411
     -7538
_____

15). 83199
     -8155
_____

16). 54666
     -5402
_____

17). 17925
     -6837
_____

18). 30301
     -3281
_____

19). 23133
     -2012
_____

20). 21153
     -3305
_____

21). 22256
     -9010
_____

22). 29653
     -9341
_____

23). 56229
     -1729
_____

24). 48038
     -7073
_____

25). 30754
     -4794
_____

26). 30324
     -8494
_____

27). 51427
     -7128
_____

28). 74660
     -4400
_____

29). 93901
     -7544
_____

30). 68315
     -4951
_____

Subtraction 5-Digit Numbers

1). 33860
-8694

2). 66804
-7062

3). 13085
-8220

4). 67985
-4124

5). 78416
-7353

6).
17744
-1874

7).
60575
-2998

8).
48608
-2906

9).
78523
-3796

10).
63390
-9381

11).
18882
-6906

12).
53269
-7835

13).
36073
-9026

14).
69386
-7169

15).
40230
-4170

16).
63437
-5456

17).
15960
-6934

18).
73372
-8960

19).
42716
-2530

20).
87443
-3416

21).
46669
-2886

22).
63582
-1188

23).
14770
-1487

24).
51367
-4399

25).
23974
-8433

26).
50668
-4519

27).
32502
-8421

28).
44506
-2145

29).
55426
-3936

30).
43577
-7947

Subtraction 5-Digit Numbers

1). 83002
    -4299
    _____

2). 45473
    -6007
    _____

3). 15038
    -1414
    _____

4). 34908
    -2651
    _____

5). 28588
    -1395
    _____

6). 47879
    -7189
    _____

7). 32944
    -1652
    _____

8). 52402
    -4921
    _____

9). 39091
    -9776
    _____

10). 55775
    -2768
    _____

11). 47390
    -2037
    _____

12). 47284
    -1364
    _____

13). 53605
    -4058
    _____

14). 14995
    -9907
    _____

15). 77826
    -9264
    _____

16). 58462
    -9127
    _____

17). 15804
    -8368
    _____

18). 71191
    -3296
    _____

19). 33742
    -3788
    _____

20). 70251
    -8108
    _____

21). 35685
    -2372
    _____

22). 65564
    -5568
    _____

23). 67297
    -6227
    _____

24). 38093
    -1833
    _____

25). 51075
    -8756
    _____

26). 14381
    -7012
    _____

27). 25774
    -5254
    _____

28). 84739
    -7556
    _____

29). 17208
    -8513
    _____

30). 63729
    -2768
    _____

Subtraction 5-Digit Numbers

1). 50676
-1223
_____

2). 25279
-6768
_____

3). 68158
-3135
_____

4). 46667
-2097
_____

5). 95838
-5369
_____

6). 46067
-6348
_____

7). 10403
-7140
_____

8). 16073
-4900
_____

9). 74720
-7899
_____

10). 91751
-5218
_____

11). 39467
-3347
_____

12). 48152
-3587
_____

13). 77857
-2283
_____

14). 64712
-2788
_____

15). 64615
-7295
_____

16). 24075
-3515
_____

17). 69804
-9574
_____

18). 11962
-3616
_____

19). 39226
-3778
_____

20). 40248
-3576
_____

21). 95704
-3995
_____

22). 57618
-4736
_____

23). 49768
-9315
_____

24). 70847
-7892
_____

25). 63992
-7428
_____

26). 75077
-1815
_____

27). 22583
-2961
_____

28). 33269
-9182
_____

29). 26663
-1991
_____

30). 93452
-8134
_____

Subtraction 5-Digit Numbers

  Time: Score: /30

1). 49339
-15445
_____

2). 46688
-30325
_____

3). 66215
-46149
_____

4). 21021
-10384
_____

5). 84672
-14840
_____

6). 46414
-13604
_____

7). 67075
-24188
_____

8). 62047
-31931
_____

9). 14231
-10371
_____

10). 61546
-42130
_____

11). 69752
-40230
_____

12). 55600
-49288
_____

13). 28525
-26057
_____

14). 31456
-28830
_____

15). 75130
-11571
_____

16). 92538
-55957
_____

17). 64488
-54400
_____

18). 77324
-17194
_____

19). 34911
-24230
_____

20). 79129
-57094
_____

21). 23264
-11915
_____

22). 47428
-34120
_____

23). 92976
-10413
_____

24). 16986
-14928
_____

25). 88314
-27224
_____

26). 63190
-54094
_____

27). 12465
-11213
_____

28). 90731
-24287
_____

29). 62774
-62343
_____

30). 68035
-28943
_____

Subtraction 5-Digit Numbers

1). 97552
-64179
_____

2). 96692
-51830
_____

3). 74251
-30289
_____

4). 27112
-14050
_____

5). 84022
-10790
_____

6). 90249
-40547
_____

7). 77835
-74738
_____

8). 49129
-26025
_____

9). 48244
-32583
_____

10). 25392
-15451
_____

11). 36112
-31990
_____

12). 85223
-72284
_____

13). 71462
-33376
_____

14). 86197
-72952
_____

15). 19922
-17529
_____

16). 95650
-14754
_____

17). 44290
-38019
_____

18). 84993
-23062
_____

19). 82812
-59651
_____

20). 17247
-15973
_____

21). 31888
-10671
_____

22). 47013
-37807
_____

23). 41165
-10588
_____

24). 23047
-18126
_____

25). 43219
-23908
_____

26). 73701
-34285
_____

27). 68345
-14032
_____

28). 44564
-13312
_____

29). 27410
-22094
_____

30). 65826
-27640
_____

Subtraction 5-Digit Numbers

Day: Time: Score: /30

1). 26011
    -22261
    ⎯⎯⎯⎯

2). 35901
    -20577
    ⎯⎯⎯⎯

3). 19351
    -15732
    ⎯⎯⎯⎯

4). 82400
    -48446
    ⎯⎯⎯⎯

5). 72173
    -47795
    ⎯⎯⎯⎯

6). 58080
    -54468
    ⎯⎯⎯⎯

7). 70976
    -35981
    ⎯⎯⎯⎯

8). 91835
    -17721
    ⎯⎯⎯⎯

9). 21846
    -13917
    ⎯⎯⎯⎯

10). 15199
     -13367
     ⎯⎯⎯⎯

11). 56840
     -23478
     ⎯⎯⎯⎯

12). 83655
     -17443
     ⎯⎯⎯⎯

13). 95649
     -39615
     ⎯⎯⎯⎯

14). 66260
     -37991
     ⎯⎯⎯⎯

15). 78853
     -41561
     ⎯⎯⎯⎯

16). 71208
     -18102
     ⎯⎯⎯⎯

17). 36475
     -20188
     ⎯⎯⎯⎯

18). 48128
     -18238
     ⎯⎯⎯⎯

19). 46999
     -16239
     ⎯⎯⎯⎯

20). 60186
     -47424
     ⎯⎯⎯⎯

21). 47540
     -46216
     ⎯⎯⎯⎯

22). 80294
     -24456
     ⎯⎯⎯⎯

23). 95425
     -66565
     ⎯⎯⎯⎯

24). 96159
     -93840
     ⎯⎯⎯⎯

25). 85089
     -72541
     ⎯⎯⎯⎯

26). 68273
     -37066
     ⎯⎯⎯⎯

27). 31902
     -10558
     ⎯⎯⎯⎯

28). 83480
     -19362
     ⎯⎯⎯⎯

29). 16294
     -12011
     ⎯⎯⎯⎯

30). 14253
     -11509
     ⎯⎯⎯⎯

Subtraction 5-Digit Numbers

1). 91037
    -80776
    _____

2). 84611
    -70074
    _____

3). 39814
    -21557
    _____

4). 70942
    -49603
    _____

5). 98032
    -49643
    _____

6). 55279
    -25662
    _____

7). 89330
    -64618
    _____

8). 84513
    -27015
    _____

9). 56594
    -16754
    _____

10). 73193
     -47679
     _____

11). 63649
     -58529
     _____

12). 95847
     -78328
     _____

13). 32525
     -19360
     _____

14). 65803
     -36560
     _____

15). 10477
     -10477
     _____

16). 63923
     -59587
     _____

17). 21721
     -19915
     _____

18). 32329
     -26527
     _____

19). 43808
     -43436
     _____

20). 98830
     -55409
     _____

21). 67428
     -32558
     _____

22). 97522
     -18341
     _____

23). 81534
     -29575
     _____

24). 59818
     -36624
     _____

25). 82704
     -26629
     _____

26). 58030
     -27734
     _____

27). 28817
     -20679
     _____

28). 30562
     -13416
     _____

29). 32336
     -20767
     _____

30). 93726
     -81037
     _____

Subtraction 5-Digit Numbers

1). 47530
   -18196
_____

2). 71916
   -58412
_____

3). 40735
   -27401
_____

4). 88799
   -21029
_____

5). 94418
   -62546
_____

6).
   42656
   -38052
_____

7).
   62648
   -26276
_____

8).
   33525
   -29369
_____

9).
   63176
   -56789
_____

10).
   95538
   -92716
_____

11).
   34732
   -29434
_____

12).
   55796
   -24272
_____

13).
   40620
   -26758
_____

14).
   85581
   -80210
_____

15).
   21993
   -19342
_____

16).
   57569
   -21391
_____

17).
   83267
   -57288
_____

18).
   65099
   -11012
_____

19).
   60334
   -13148
_____

20).
   76000
   -65555
_____

21).
   69454
   -63490
_____

22).
   64010
   -35844
_____

23).
   81006
   -30443
_____

24).
   35392
   -33106
_____

25).
   77653
   -70515
_____

26).
   66992
   -23709
_____

27).
   80875
   -58470
_____

28).
   14698
   -13545
_____

29).
   83344
   -28625
_____

30).
   88017
   -69890
_____

Multiplication X

20 worksheets

30 problems per sheet

William. Education

Examples

Example 1 (2-digit by 1-digit numbers)

1
$$
\begin{array}{r}
95 \\
\times\ \ 5 \\
\hline
\end{array}
$$

2
$$
\begin{array}{r}
^{2}95 \\
\times\ \ 5 \\
\hline
5 \\
\end{array}
$$

3
$$
\begin{array}{r}
^{2}95 \\
\times\ \ 5 \\
\hline
475 \\
\end{array}
$$

Example 2 (3-digit by 1-digit numbers)

1
$$
\begin{array}{r}
346 \\
\times\ \ 8 \\
\hline
\end{array}
$$

2
$$
\begin{array}{r}
^{4}346 \\
\times\ \ 8 \\
\hline
8 \\
\end{array}
$$

3
$$
\begin{array}{r}
^{3\ 4}346 \\
\times\ \ 8 \\
\hline
68 \\
\end{array}
$$

4
$$
\begin{array}{r}
^{3\ 4}346 \\
\times\ \ 8 \\
\hline
2768 \\
\end{array}
$$

Example 3 (4-digit by 1-digit numbers)

1
$$
\begin{array}{r}
2435 \\
\times\ \ 6 \\
\hline
\end{array}
$$

2
$$
\begin{array}{r}
^{3}2435 \\
\times\ \ 6 \\
\hline
0 \\
\end{array}
$$

3
$$
\begin{array}{r}
^{2\ 3}2435 \\
\times\ \ 6 \\
\hline
10 \\
\end{array}
$$

4
$$
\begin{array}{r}
^{2\ 2\ 4}2435 \\
\times\ \ 6 \\
\hline
610 \\
\end{array}
$$

5
$$
\begin{array}{r}
^{2\ 2\ 4}2435 \\
\times\ \ 6 \\
\hline
14610 \\
\end{array}
$$

Example 4 (5-digit by 1-digit numbers)

1
$$
\begin{array}{r}
42563 \\
\times\ \ 6 \\
\hline
\end{array}
$$

2
$$
\begin{array}{r}
^{1}42563 \\
\times\ \ 6 \\
\hline
8 \\
\end{array}
$$

3
$$
\begin{array}{r}
^{3\ 1}42563 \\
\times\ \ 6 \\
\hline
78 \\
\end{array}
$$

4
$$
\begin{array}{r}
^{3\ 3\ 1}42563 \\
\times\ \ 6 \\
\hline
378 \\
\end{array}
$$

5+
$$
\begin{array}{r}
^{1\ 3\ 3\ 1}42563 \\
\times\ \ 6 \\
\hline
255378 \\
\end{array}
$$

Example 5 (3-digit by 3-digit numbers)

1
```
    247
  x
    528
```

2
```
      5
    247
  x   ↑
    528
      6
```

3
```
    3 5
    247
  x   ↑
    528
     76
```

4
```
    3 5
    247
  x   ↖
    528
   1976
```

5
```
    3 5
    247
  x
    528
   1976
      0
```

6
```
      1
    3 5
    247
  x   ↑
    528
   1976
     40
```

7
```
    0 1
    3 5
    247
  x   ↑
    528
   1976
    940
```

8
```
    0 1
    3 5
    247
  x   ↖
    528
   1976
   4940
```

9
```
    3 5
    247
  x
    528
   1976
   4940
     00
```

10
```
      3
    0 1
    3 5
    247
  x   ↗
    528
   1976
     40
    500
```

11
```
    2 3
    0 1
    3 5
    247
  x   ↗
    528
   1976
    940
   3500
```

12
```
    2 3
    0 1
    3 5
    247
  x   ↑
    528
   1976
   4940
 123500
```

13
```
    2 3
    0 1
    3 5
    247
  x
    528
   1 2 1
   1976
   4940
 +123500
 130416
```

Example 6 (2-digit by 2-digit numbers)

1
```
   45
 x
   95
_____
```

2
```
   2
   45
 x  ↑
   95
_____
    5
```

3
```
   2
   45
 x  ↖
   95
_____
  225
```

4
```
   2
   45
 x
   95
_____
  225
    0
```

5
```
   4
   2
   45
 x  ↗
   95
_____
  225
   50
```

6
```
   4
   2
   45
 x  ↑
   95
_____
  225
 4050
```

7
```
     4
     2
   x 45
     95
  _____
     225
 + 4050
  _____
    4275
```

Division 5-Digit By 1-Digit Numbers

Day: ⏱ **Time:**   **Score:**   **/40**

7)14595	3)18808	5)82213	1)89277	5)21985
6)55202	3)31109	5)86902	6)41601	8)77806
2)22255	2)70291	5)25033	2)83691	7)41808
5)61133	4)96406	1)83549	8)78218	8)51888
3)48230	3)30127	4)77202	3)78451	9)69974
7)87813	2)54861	7)87197	6)24352	7)75971
8)51484	7)35830	4)18265	8)24373	8)10451
9)13791	2)49529	4)18840	8)65835	7)74163

Division 5-Digit By 1-Digit Numbers

Day: ⏱ **Time:**   **Score:**   /40

6)62510	1)41007	8)28749	8)97540	2)11853
4)82046	3)60381	7)50754	2)54275	7)87159
1)31783	2)66737	1)76857	7)68120	4)32512
3)52945	7)98567	7)37934	7)10616	5)38948
8)40737	9)69378	4)18940	1)59118	7)89076
4)63103	2)20882	7)92989	8)54651	5)85613
7)90782	5)39216	2)79062	9)41686	7)43053
5)65787	2)31884	3)27653	7)62991	9)18259

Day:　　　🕐 **Time:**　　　**Score:**　　/40

6⟌55713	7⟌47232	4⟌39044	2⟌29920	1⟌62158
7⟌76388	9⟌79949	7⟌63054	5⟌68217	3⟌73866
8⟌92928	9⟌95678	2⟌42738	9⟌94179	8⟌80002
1⟌34043	9⟌38155	8⟌41395	5⟌18774	8⟌43605
4⟌84044	2⟌94463	8⟌11761	8⟌16393	5⟌43743
7⟌90794	6⟌92101	9⟌84859	7⟌43794	6⟌95773
6⟌83833	2⟌39603	7⟌65257	7⟌83111	3⟌64379
5⟌47122	3⟌76670	8⟌62997	3⟌52171	1⟌81584

Division 5-Digit By 1-Digit Numbers

Day: ⏱ **Time:**   **Score:**   /40

8⟌95966	8⟌78550	7⟌11985	2⟌18341	5⟌61996
8⟌97884	8⟌47696	1⟌25335	6⟌77913	7⟌57328
4⟌34271	8⟌19865	8⟌53822	6⟌19460	2⟌86311
2⟌35741	5⟌92310	3⟌57495	6⟌46657	2⟌63186
1⟌80666	7⟌80736	4⟌29440	7⟌85886	5⟌90447
8⟌64092	5⟌26517	1⟌93521	3⟌36094	4⟌61821
4⟌52916	7⟌61929	1⟌14633	4⟌12540	8⟌78519
2⟌12244	2⟌23810	7⟌56747	6⟌24398	4⟌61847

Division 5-Digit By 1-Digit Numbers

Day: ⏱ Time: Score: /40

2⟌25719	7⟌20097	6⟌18533	8⟌10797	3⟌70269
3⟌62142	7⟌99338	4⟌18731	6⟌39081	4⟌55647
4⟌82178	5⟌99028	6⟌12601	7⟌88804	7⟌28492
3⟌49532	5⟌30702	8⟌24660	1⟌28831	4⟌82974
8⟌21740	2⟌37527	5⟌28022	3⟌17758	7⟌66203
8⟌20820	2⟌44272	4⟌16147	8⟌15955	1⟌56718
7⟌90869	2⟌75624	1⟌89824	4⟌68872	4⟌53967
3⟌26248	5⟌48368	5⟌64790	5⟌91813	4⟌53134

Division 6-Digit By 1-Digit Numbers

6)700712   2)213983   1)960694   7)373964   5)584830

2)272484   4)798869   4)651854   4)296690   2)552624

8)795414   5)660713   6)301269   1)902630   3)654003

8)617809   1)526046   9)315294   3)638135   7)299837

6)776751   1)900555   5)170344   2)283966   2)584466

4)399695   6)212258   1)519459   4)571056   6)651364

1)422655   5)467003   1)139818   8)579484   7)671877

4)976687   8)365227   1)404889   3)168730   8)830356

Division 6-Digit By 1-Digit Numbers

3⟌955851   3⟌677366   5⟌474761   2⟌120677   8⟌870610

7⟌430602   3⟌756891   5⟌172314   8⟌245283   4⟌701897

4⟌721382   6⟌630067   5⟌722557   5⟌244933   3⟌467518

8⟌546609   3⟌164716   3⟌430936   3⟌854161   1⟌329333

7⟌706410   3⟌428855   1⟌277655   3⟌309058   6⟌263760

9⟌699500   6⟌795817   7⟌178352   6⟌809374   6⟌954949

2⟌107937   6⟌549017   2⟌535132   2⟌148975   5⟌298771

8⟌591363   6⟌633734   2⟌900817   4⟌499662   4⟌902574

Division 6-Digit By 1-Digit Numbers

6|744614   7|143281   7|451822   4|527056   2|931185

3|115421   7|941537   4|389529   8|527794   6|798501

2|252306   9|970812   2|889823   7|571693   3|146509

9|303628   7|676559   8|957065   4|595922   5|874339

3|132283   6|788845   2|227381   9|975811   6|294508

1|694093   8|176583   2|989915   4|592290   8|563426

1|467699   5|593679   2|144067   2|264496   3|119254

3|905867   2|207885   2|446567   2|705816   8|584562

Division 6-Digit By 1-Digit Numbers

Day: ⏱ **Time:**   **Score:**   **/40**

7)928048	4)224710	2)262185	7)212958	4)682544
9)754258	9)584998	7)295520	5)821708	7)178337
5)496836	4)263563	4)841276	4)338152	6)541396
6)571294	6)412826	5)349277	9)646184	6)381310
7)616589	6)164004	5)426622	8)436357	2)924470
5)978241	5)810601	2)767269	7)171154	9)335743
6)581258	4)872640	5)432274	8)916154	7)902252
7)637771	6)906341	5)648388	3)603962	1)763836

Division 6-Digit By 1-Digit Numbers

Day: ⏱ **Time:** **Score:** /40

5)415712	2)390306	1)164828	8)145065	1)477862
6)466465	7)162936	2)999782	4)145131	5)783632
1)527612	3)635682	5)912730	4)328546	3)607605
2)494503	2)663626	7)776228	1)393363	3)769198
3)691621	5)944993	7)302156	7)504519	2)309008
3)648936	1)756091	2)921163	6)573708	2)622089
7)502213	6)137581	6)186136	3)781158	8)475237
3)704640	1)318240	4)708323	3)601513	6)881253

Division 4-Digit By 2-Digit Numbers

Day: ⏱ Time: Score: /40

43⟌3601	30⟌9405	35⟌7265	14⟌3978	17⟌2082
48⟌9572	78⟌1307	90⟌2248	68⟌6296	95⟌8238
48⟌9401	57⟌5751	83⟌5975	83⟌5194	14⟌4896
73⟌3302	85⟌5743	29⟌5282	86⟌4972	28⟌1911
37⟌8968	20⟌8095	23⟌6734	94⟌8617	43⟌1870
41⟌4768	87⟌9787	77⟌6189	43⟌6602	79⟌4737
22⟌8856	93⟌5949	22⟌2401	58⟌9881	56⟌6025
94⟌8203	54⟌7586	89⟌9707	15⟌6793	55⟌3881

Division 4-Digit By 2-Digit Numbers

67⟌9884	91⟌4409	93⟌9160	92⟌5199	66⟌1079
23⟌6017	97⟌7929	93⟌5941	99⟌9041	34⟌7214
63⟌3033	86⟌1671	82⟌4214	65⟌5878	68⟌6900
81⟌4945	88⟌4362	22⟌3175	44⟌2405	72⟌4858
15⟌5438	55⟌4379	61⟌9129	27⟌4301	11⟌6745
93⟌3449	69⟌8817	60⟌9648	98⟌9615	95⟌1979
76⟌1045	37⟌3437	78⟌2570	45⟌7806	23⟌8093
45⟌4065	38⟌2553	23⟌2742	60⟌4748	52⟌9378

Division 4-Digit By 2-Digit Numbers

Day: ⏱ **Time:**   **Score:**   /40

29⟌5412	18⟌7719	99⟌8576	20⟌8285	80⟌9193
23⟌3904	35⟌6011	95⟌6146	89⟌3106	68⟌9160
11⟌6148	48⟌2504	71⟌9284	94⟌1249	25⟌1491
97⟌9139	81⟌6972	76⟌6559	61⟌3914	52⟌8039
68⟌2035	22⟌4409	83⟌2601	70⟌2126	85⟌8444
30⟌4744	91⟌3860	36⟌7803	87⟌4714	37⟌4757
53⟌2912	14⟌4075	15⟌7827	75⟌8813	18⟌7343
24⟌3149	78⟌7688	70⟌1781	37⟌5427	91⟌5060

Division 4-Digit By 2-Digit Numbers

Day: ⏱ **Time:**   **Score:**   /40

56)5175   73)7271   83)3747   45)8870   56)9988

66)4077   53)4352   67)7076   76)3040   17)7149

48)3531   37)3527   45)9817   50)7512   23)8009

86)8361   78)5299   85)5617   80)7854   39)1479

85)4970   39)2000   32)1714   34)6939   34)3904

23)9985   41)9062   51)6630   68)9168   73)5447

36)8634   57)2036   62)2357   56)2061   99)6038

82)9431   88)3121   37)6070   17)1261   81)1872

Division 4-Digit By 2-Digit Numbers

51 | 7808　　73 | 4847　　36 | 2311　　96 | 4456　　39 | 4763

57 | 6119　　22 | 9166　　66 | 5242　　32 | 8625　　55 | 5033

43 | 1536　　91 | 7218　　41 | 4534　　26 | 3644　　77 | 6986

81 | 2000　　36 | 1659　　48 | 7168　　34 | 9871　　24 | 1877

39 | 8739　　78 | 3151　　60 | 2517　　61 | 6096　　38 | 7923

30 | 8838　　44 | 6738　　70 | 7898　　15 | 1773　　28 | 7589

20 | 4344　　50 | 4016　　17 | 6042　　29 | 3107　　69 | 2230

58 | 9849　　91 | 5019　　14 | 8528　　64 | 3575　　22 | 2660

Division 5-Digit By 2-Digit Numbers

Day: **Time:**   **Score:**   /40

79 | 84764   61 | 43101   17 | 85644   24 | 62100   74 | 96520

69 | 91620   23 | 11482   21 | 18247   57 | 93043   76 | 12226

39 | 34924   73 | 26736   47 | 14436   60 | 25931   29 | 84814

70 | 45126   86 | 79558   42 | 47292   96 | 84109   97 | 46076

59 | 35882   99 | 60659   56 | 94869   80 | 38161   59 | 33776

38 | 94724   39 | 38037   18 | 22089   29 | 19385   69 | 42179

33 | 14039   20 | 93863   63 | 14006   33 | 42648   37 | 41194

54 | 91345   32 | 75358   30 | 72121   41 | 42341   13 | 86302

Division 5-Digit By 2-Digit Numbers

87)90899　　52)79837　　34)17210　　38)60529　　31)11566

79)63668　　15)99679　　80)87522　　56)59257　　47)50912

97)11906　　12)66523　　99)28542　　28)92638　　70)81312

52)79086　　56)23622　　35)46132　　88)96821　　63)11639

9)32558　　46)26110　　80)79612　　69)41008　　10)44953

84)60762　　63)39783　　81)84860　　85)61567　　69)65119

18)44130　　48)58155　　44)77765　　9)70947　　54)45421

13)87571　　44)65566　　86)63274　　41)76697　　59)18065

Division 5-Digit By 2-Digit Numbers

60⟌27929　　79⟌70764　　81⟌24150　　44⟌30470　　97⟌46964

88⟌99663　　11⟌79559　　77⟌73468　　20⟌86652　　77⟌22441

87⟌67931　　43⟌79244　　49⟌66017　　42⟌33938　　62⟌44671

76⟌40819　　92⟌83080　　69⟌13098　　93⟌12956　　67⟌78797

81⟌40658　　62⟌41869　　33⟌88430　　26⟌55172　　98⟌16401

94⟌70689　　56⟌80990　　89⟌27906　　96⟌62695　　62⟌77619

88⟌60042　　87⟌98301　　24⟌38534　　86⟌68820　　53⟌96800

84⟌17182　　77⟌81276　　49⟌55487　　77⟌59121　　85⟌51507

Division 5-Digit By 2-Digit Numbers

Day: ⏱ **Time:**   **Score:**   /40

21)25489   73)26325   62)75535   67)38871   27)76341

33)30869   52)53441   23)74917   55)87684   28)46635

12)39988   32)76641   71)34347   46)48090   68)65880

58)99720   95)42473   57)96747   51)60092   70)76784

51)90583   18)46130   71)24256   78)44686   58)72703

65)19580   75)12385   31)38856   43)81652   57)78056

95)80630   40)59789   23)31451   89)35040   82)46953

27)61997   62)15049   95)10722   93)22294   12)22640

Division 5-Digit By 2-Digit Numbers

44|91540   66|47051   92|22828   61|86245   48|70788

40|83532   78|76139   12|26734   82|65587   92|10818

60|21359   53|96524   65|14856   15|31577   81|49538

31|39791   42|61604   44|36128   56|75563   9|21283

87|79657   42|30947   10|86405   61|26096   31|66074

98|38793   20|40415   94|26552   36|41147   66|33992

63|72306   79|96867   56|92434   96|24129   18|62108

75|20314   16|61725   79|68436   14|42651   74|12896

Division ÷

20 worksheets
40 problems per sheet

William. Education

Examples

Quotient

Divisor | **Dividend**

Dividend ÷ Divisor = Quotient

Steps

D = Divide
M = Multiply
S = Subtract
B = Bring down

Example :

```
7 | 917
```

```
    1
7 | 917
```

Step 1: D for Divide

How many times will 7 go into 917? That's too hard to work out in your head, so let's break it down into smaller steps.
The first problem you'll work out in this equation is how many times can you divide 9 into 7. The answer is 1. So you put 1 on the quotient line.

```
    1
7 | 917
    7
```

Step 2: M for Multiply

You multiply your answer from step 1 and your divisor: 1 x 7 = 7. You write 7 under the 9

```
    1
7 | 917
    7
    2
```

Step 3: S for Subtract

Next you subtract. In this case it will be 9 – 7 = 2

```
    1
7 | 917
    7
   21
```

Step 4: B for Bring down

The last step in the sequence is to bring down the next number from the dividend, which in this case is 1. You write the 1 next to the 2, making the number 21.

Now you start all over again

```
   13
7 | 917
    7
   21
   21
```

Step 1: D for Divide

How many times can you divide 3into 21. The answer is 3. So you put 3 on the quotient line.

```
   13
7 | 917
    7
   21
   21
```

Step 2: M for Multiply

You multiply your answer from step 1 and your divisor: 3 x 7 = 21. You write 21 under the 21

```
      13
  7 | 917
      7
    ────
     21
     21
    ────
      0
```

Next you subtract. In this case it will be 21 – 21 = 0

```
      13
  7 | 917
      7
    ────
     21
     21
    ────
     07
```

The last step in the sequence is to bring down the next number from the dividend, which in this case is 7. You write the 7 next to the 0, making the number 7.

Now you start all over again

```
      131
  7 | 917
      7
    ────
     21
     21
    ────
     07
```

How many times can you divide 7 into 7. The answer is 1. So you put 1 on the quotient line.

```
      131
  7 | 917
      7
    ────
     21
     21
    ────
     07
      7
    ────
```

You multiply your answer from step 1 and your divisor: 1 x 7 = 7. You write 7 under the 7

```
      131
  7 | 917
      7
    ────
     21
     21
    ────
     07
      7
    ────
      0
```

Next you subtract. In this case it will be 7– 7 = 0

There is no need for step 4. We l finished the problem.

Once you have the answer, do tl problem in reverse using multiplica (7 x 131 = 917) to make sure yo answer is correct.

Division 5-digit by 2

```
        00309
  88 | 27218
       - 0
      ─────
        27
       - 0
      ─────
        272
      - 264
      ─────
         81
        - 0
      ─────
        818
      - 792
      ─────
         26
```

27218 divided by 88 equals 309 with a remainder of 26

Division 3-digit by 2

```
         007
  45 | 323
       - 0
      ─────
        32
       - 0
      ─────
        323
      - 315
      ─────
          8
```

323 divided by 45 equals 7 with a remainder of 8

Division 4-digit by 1

```
        0928
  8 | 7426
      - 0
     ─────
       74
      - 72
     ─────
        22
      - 16
     ─────
        66
      - 64
     ─────
         2
```

7426 divided by 8 equals 928 with a remainder of 2

Division 4-digit by 2

```
        0132
  32 | 4236
       - 0
      ─────
        42
      - 32
      ─────
        103
       - 96
      ─────
         76
       - 64
      ─────
         12
```

4236 divided by 32 equals 132 with a remainder of 12

Division 5-digit by 1

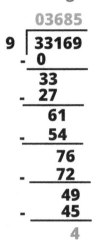

```
      03685
  9 | 33169
    -  0
      ───
      33
    - 27
      ───
      61
    - 54
      ───
      76
    - 72
      ───
      49
    - 45
      ───
       4
```

33169 divided by 9 equals
3685 with a remainder of 4

Once you have the
answer, do the
problem in reverse
using multiplication
(9 x 3685+4 = 33169)
to make sure your
answer is correct.

Division 3-digit by 1

```
        092
  7 |  648
    -   0
      ────
       64
    -  63
      ────
       18
    -  14
      ────
        4
```

648 divided by 7 equals
92 with a remainder of 4

Once you have the
answer, do the
problem in reverse
using multiplication
(7 x 92+4 = 648)
to make sure your
answer is correct.

Division 4-digit by 1

```
       0553
  5 |  2767
    -   0
      ────
       27
    - 25
      ────
       26
    - 25
      ────
       17
    - 15
      ────
        2
```

2767 divided by 5 equals
553 with a remainder of 2

Once you have the
answer, do the
problem in reverse
using multiplication
(5 x 553+2 = 2767)
to make sure your
answer is correct.

Division 6-digit by 1

```
      065282
  7 | 456978
    -  0
      ─────
      45
    - 42
      ─────
      36
    - 35
      ─────
      19
    - 14
      ─────
      57
    - 56
      ─────
      18
    - 14
      ─────
       4
```

456978 divided by 7 equals
65282 with a remainder of 4

Once you have the answer, do the
problem in reverse using
multiplication (7 x 65282+4 = 456978)
to make sure your answer is correct.

Division 5-digit by 2

```
       02883
 12 | 34598
    -   0
      ─────
      34
    - 24
      ─────
      105
    -  96
      ─────
       99
    -  96
      ─────
       38
    -  36
      ─────
        2
```

34598 divided by 12 equals
2883 with a remainder of 2

Once you have the
answer, do the
problem in reverse
using multiplication
(12 x 2883+2 = 34598)
to make sure your
answer is correct.

Division 3-digit by 2

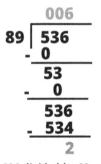

```
         006
 89 |  536
    -   0
      ────
       53
    -   0
      ────
      536
    - 534
      ────
        2
```

536 divided by 89 equals
6 with a remainder of 2

Once you have the
answer, do the
problem in reverse
using multiplication
(89 x 6+2 = 536)
to make sure your
answer is correct.

Division 4-digit by 1

```
        0949
  6 |  5698
    -   0
      ────
       56
    -  54
      ────
       29
    -  24
      ────
       58
    -  54
      ────
        4
```

5698 divided by 6 equals
949 with a remainder of 4

Once you have the
answer, do the
problem in reverse
using multiplication
(949 x 6+4 = 5698)
to make sure your
answer is correct.

Division 4-digit by 2

```
        0179
 19 | 3417
    -   0
      ─────
      34
    - 19
      ─────
      151
    - 133
      ─────
      187
    - 171
      ─────
       16
```

3417 divided by 19 equals
179 with a remainder of 16

Once you have the answer, do the
problem in reverse using
multiplication (179 x 19+16 = 3417)
to make sure your answer is correct.

 **Once you have the answer, do the problem in reverse using multiplication
(Quotient x Divisor+Remainder = Dividend) to make sure your answer is correct.**

Multiplication 4-Digit By 1-Digit Numbers

Day: Time: Score: /30

1).　6323
　　　　x7
_____

2).　4203
　　　　x9
_____

3).　1842
　　　　x6
_____

4).　6080
　　　　x8
_____

5).　1340
　　　　x2
_____

6).
　4856
　　x9
_____

7).
　5942
　　x8
_____

8).
　9622
　　x8
_____

9).
　2578
　　x7
_____

10).
　5939
　　x5
_____

11).
　3569
　　x1
_____

12).
　6430
　　x9
_____

13).
　3234
　　x6
_____

14).
　9175
　　x4
_____

15).
　2345
　　x8
_____

16).
　2651
　　x8
_____

17).
　1269
　　x9
_____

18).
　9087
　　x7
_____

19).
　1188
　　x1
_____

20).
　8174
　　x3
_____

21).
　3642
　　x8
_____

22).
　6075
　　x4
_____

23).
　4954
　　x6
_____

24).
　7505
　　x9
_____

25).
　7819
　　x8
_____

26).
　2498
　　x4
_____

27).
　9418
　　x8
_____

28).
　7119
　　x2
_____

29).
　7980
　　x8
_____

30).
　5009
　　x5
_____

Multiplication 4-Digit By 1-Digit Numbers

Day: Time: Score: /30

1).   1378
      x9
_____

2).   5252
      x2
_____

3).   5676
      x2
_____

4).   6630
      x4
_____

5).   1746
      x7
_____

6).   3549
      x2
_____

7).   6231
      x8
_____

8).   1644
      x7
_____

9).   1382
      x5
_____

10).   8793
      x3
_____

11).   1660
      x1
_____

12).   8986
      x8
_____

13).   4416
      x5
_____

14).   2620
      x2
_____

15).   5491
      x2
_____

16).   5814
      x4
_____

17).   7875
      x3
_____

18).   4881
      x9
_____

19).   8590
      x2
_____

20).   9795
      x8
_____

21).   9596
      x8
_____

22).   2342
      x1
_____

23).   9848
      x2
_____

24).   8663
      x3
_____

25).   3574
      x4
_____

26).   5066
      x8
_____

27).   9492
      x4
_____

28).   5305
      x2
_____

29).   5945
      x2
_____

30).   2002
      x6
_____

Multiplication 4-Digit By 1-Digit Numbers

Day: **Time:** **Score:** **/30**

1).   4079
     x8

2).   3534
     x3

3).   3736
     x2

4).   7953
     x7

5).   2464
     x6

6).   1034
     x2

7).   9522
     x1

8).   8468
     x5

9).   8365
     x8

10).   8659
     x7

11).   6926
     x2

12).   6519
     x3

13).   9448
     x4

14).   3939
     x7

15).   5900
     x5

16).   9985
     x3

17).   6824
     x1

18).   8910
     x2

19).   5969
     x3

20).   1274
     x6

21).   7240
     x8

22).   6812
     x8

23).   6260
     x2

24).   5210
     x9

25).   6350
     x3

26).   3629
     x3

27).   8823
     x3

28).   1604
     x6

29).   6287
     x2

30).   5985
     x5

Multiplication 4-Digit By 1-Digit Numbers

Day: **Time:** **Score:** /30

1). 9119
x6

2). 6967
x8

3). 9897
x1

4). 5018
x2

5). 6609
x9

6). 6181
x7

7). 7314
x5

8). 8469
x3

9). 1463
x6

10). 3011
x4

11). 1876
x8

12). 7597
x8

13). 8593
x8

14). 3001
x8

15). 6830
x3

16). 1703
x1

17). 4428
x2

18). 9106
x8

19). 4153
x7

20). 6116
x4

21). 8120
x5

22). 2230
x2

23). 8444
x8

24). 3934
x6

25). 5944
x1

26). 8206
x8

27). 3331
x7

28). 1456
x5

29). 9884
x5

30). 5031
x1

Multiplication 4-Digit By 1-Digit Numbers

Day: **Time:** **Score:** **/30**

1). 8929
   x5

2). 1876
   x6

3). 8489
   x3

4). 6829
   x7

5). 3237
   x4

6). 6185
   x6

7). 8939
   x1

8). 1953
   x3

9). 5189
   x8

10). 2037
   x6

11). 1816
   x5

12). 3745
   x1

13). 8297
   x2

14). 8193
   x1

15). 3377
   x7

16). 9760
   x1

17). 4580
   x1

18). 8064
   x7

19). 3723
   x6

20). 5672
   x6

21). 7863
   x4

22). 6908
   x6

23). 4314
   x7

24). 5668
   x4

25). 9517
   x5

26). 5117
   x7

27). 6582
   x2

28). 6494
   x6

29). 1246
   x2

30). 7038
   x5

Multiplication 5-Digit By 1-Digit Numbers

Day: ⏱ Time: Score: /30

1). 61647
      x4

2). 47851
      x4

3). 86829
      x8

4). 15525
      x5

5). 64700
      x7

6). 34907
      x3

7). 66354
      x6

8). 66413
      x8

9). 48166
      x6

10). 16919
      x1

11). 65129
      x7

12). 41676
      x8

13). 89009
      x7

14). 42430
      x7

15). 13941
      x4

16). 38619
      x6

17). 92874
      x6

18). 90151
      x5

19). 13194
      x7

20). 10910
      x7

21). 88799
      x1

22). 11902
      x5

23). 98910
      x7

24). 51736
      x6

25). 21002
      x4

26). 20008
      x7

27). 66363
      x5

28). 86053
      x3

29). 51996
      x1

30). 61560
      x4

Multiplication 5-Digit By 1-Digit Numbers

Day: 🕙 **Time:**   **Score:**   **/30**

1). 13343
x1

2). 13780
x3

3). 47376
x3

4). 77288
x9

5). 63734
x6

6). 90990
x6

7). 84611
x3

8). 90652
x1

9). 84672
x3

10). 15136
x6

11). 28788
x7

12). 99176
x1

13). 70537
x7

14). 78850
x2

15). 78401
x2

16). 93279
x4

17). 58887
x7

18). 92892
x5

19). 39683
x6

20). 48029
x8

21). 91406
x5

22). 41601
x8

23). 33489
x4

24). 26084
x2

25). 18333
x2

26). 28207
x7

27). 57957
x5

28). 14341
x2

29). 79089
x9

30). 65846
x1

Multiplication 5-Digit By 1-Digit Numbers

Day:　　　**Time:**　　　**Score:**　　**/30**

1). 30891
　　x5

2). 61783
　　x5

3). 45720
　　x4

4). 18563
　　x5

5). 45248
　　x5

6). 57559
　　x6

7). 17437
　　x5

8). 50201
　　x5

9). 77922
　　x2

10). 32701
　　x2

11). 19093
　　x6

12). 21702
　　x8

13). 58545
　　x4

14). 49419
　　x6

15). 95832
　　x2

16). 60398
　　x5

17). 97601
　　x7

18). 28941
　　x3

19). 16389
　　x8

20). 82225
　　x3

21). 17634
　　x7

22). 79201
　　x9

23). 32778
　　x4

24). 69470
　　x3

25). 53738
　　x3

26). 49496
　　x5

27). 45462
　　x4

28). 65764
　　x2

29). 58837
　　x7

30). 85783
　　x2

48

Multiplication 5-Digit By 1-Digit Numbers

Day: **Time:** **Score:** **/30**

1). 39642
   x6

2). 78930
   x4

3). 98602
   x2

4). 61014
   x5

5). 10031
   x6

6). 31206
   x5

7). 50454
   x9

8). 71514
   x8

9). 66260
   x8

10). 17123
   x8

11). 62869
   x9

12). 19140
   x3

13). 73046
   x9

14). 67067
   x4

15). 75222
   x5

16). 63330
   x8

17). 72886
   x7

18). 73175
   x7

19). 70377
   x2

20). 83306
   x4

21). 38739
   x3

22). 81721
   x7

23). 67697
   x5

24). 19412
   x8

25). 26249
   x3

26). 91803
   x8

27). 74918
   x8

28). 14119
   x5

29). 97257
   x1

30). 20217
   x5

Multiplication 5-Digit By 1-Digit Numbers

Day: Time: Score: /30

1). 74197
    x9

2). 53518
    x1

3). 86181
    x9

4). 13386
    x6

5). 72959
    x7

6). 99867
    x1

7). 99210
    x6

8). 61068
    x5

9). 44864
    x2

10). 65025
    x7

11). 67509
    x3

12). 20822
    x3

13). 84791
    x2

14). 64814
    x1

15). 99530
    x3

16). 91403
    x3

17). 15378
    x4

18). 82497
    x7

19). 77546
    x6

20). 60970
    x7

21). 61399
    x3

22). 11249
    x7

23). 51484
    x6

24). 44485
    x8

25). 14992
    x8

26). 66685
    x2

27). 76537
    x5

28). 15635
    x5

29). 85691
    x5

30). 66504
    x7

Multiplication 4-Digit By 2-Digit Numbers

Day: ⏱ **Time:**   **Score:**   /30

1). 5114
   x94

2). 2391
   x64

3). 9113
   x94

4). 7621
   x58

5). 2713
   x16

6). 9741
   x53

7). 1146
   x36

8). 4059
   x83

9). 5980
   x65

10). 9988
   x32

11). 3394
   x16

12). 6515
   x73

13). 8544
   x83

14). 3627
   x34

15). 4813
   x88

16). 1770
   x26

17). 3977
   x86

18). 8726
   x82

19). 4039
   x51

20). 4572
   x63

21). 1525
   x48

22). 3539
   x88

23). 2544
   x76

24). 4654
   x21

25). 8821
   x44

26). 1022
   x97

27). 6183
   x42

28). 7491
   x94

29). 8725
   x60

30). 6223
   x44

Multiplication 4-Digit By 2-Digit Numbers

Day: 🕐 **Time:**   **Score:**   **/30**

1).   6428
    x91

2).   8789
    x18

3).   3133
    x86

4).   5643
    x74

5).   3093
    x35

6).   1240
    x69

7).   4315
    x66

8).   9149
    x74

9).   6474
    x86

10).   5478
    x63

11).   2233
    x67

12).   8839
    x32

13).   5154
    x15

14).   8622
    x67

15).   1751
    x89

16).   6234
    x36

17).   7958
    x65

18).   7322
    x70

19).   4592
    x56

20).   5549
    x72

21).   6412
    x83

22).   2222
    x40

23).   4016
    x22

24).   9868
    x45

25).   5128
    x18

26).   7934
    x15

27).   1803
    x32

28).   8894
    x47

29).   4688
    x33

30).   6427
    x22

Multiplication 4-Digit By 2-Digit Numbers

Day: ⏱ **Time:**   **Score:**   **/30**

1). 3274
    x44
_____

2). 9601
    x55
_____

3). 9663
    x24
_____

4). 8228
    x41
_____

5). 7263
    x58
_____

6).
    5849
    x58
_____

7).
    4409
    x91
_____

8).
    1150
    x26
_____

9).
    4221
    x27
_____

10).
    5861
    x84
_____

11).
    6992
    x90
_____

12).
    5651
    x27
_____

13).
    1050
    x27
_____

14).
    4133
    x83
_____

15).
    3375
    x14
_____

16).
    6360
    x62
_____

17).
    9942
    x41
_____

18).
    1502
    x65
_____

19).
    1951
    x48
_____

20).
    6661
    x38
_____

21).
    9618
    x27
_____

22).
    7860
    x97
_____

23).
    2204
    x20
_____

24).
    8308
    x37
_____

25).
    5267
    x98
_____

26).
    8986
    x61
_____

27).
    3954
    x38
_____

28).
    6428
    x29
_____

29).
    6266
    x85
_____

30).
    7347
    x44
_____

Multiplication 4-Digit By 2-Digit Numbers

Day: 🕑 **Time:**   **Score:**   /30

1).   4708
     x48
_____

2).   1723
     x15
_____

3).   4275
     x62
_____

4).   3918
     x54
_____

5).   6489
     x93
_____

6).   5993
     x65
_____

7).   7417
     x89
_____

8).   5097
     x56
_____

9).   6327
     x73
_____

10).   1694
     x88
_____

11).   9756
     x21
_____

12).   9186
     x35
_____

13).   9658
     x9
_____

14).   9700
     x22
_____

15).   9456
     x82
_____

16).   2056
     x70
_____

17).   7759
     x80
_____

18).   7577
     x38
_____

19).   4099
     x10
_____

20).   5203
     x55
_____

21).   6287
     x72
_____

22).   7313
     x79
_____

23).   9885
     x61
_____

24).   6768
     x25
_____

25).   4877
     x73
_____

26).   6955
     x36
_____

27).   8780
     x39
_____

28).   8016
     x48
_____

29).   1394
     x33
_____

30).   5048
     x17
_____

Multiplication 4-Digit By 2-Digit Numbers

Day: 🕐 **Time:**   **Score:**   **/30**

1).   4457
      x26

2).   6110
      x40

3).   8459
      x89

4).   2769
      x57

5).   6323
      x37

6).   3393
      x43

7).   5727
      x44

8).   8561
      x81

9).   8334
      x90

10).   8916
       x92

11).   7231
       x41

12).   9176
       x22

13).   5758
       x42

14).   5200
       x95

15).   3433
       x92

16).   3064
       x28

17).   1517
       x93

18).   3261
       x60

19).   4500
       x45

20).   5729
       x32

21).   4494
       x51

22).   9398
       x60

23).   6027
       x36

24).   7341
       x32

25).   9267
       x71

26).   7417
       x83

27).   6030
       x48

28).   3514
       x34

29).   5258
       x63

30).   7080
       x76

Multiplication 5-Digit By 3-Digit Numbers

Day: 🕑 **Time:**   **Score:**   **/30**

1).   18395
     x158

2). 85447
     x828

3). 61519
     x651

4). 99979
     x100

5). 72830
     x895

6).
    34233
     x379

7).
    73273
     x492

8).
    27148
     x654

9).
    43126
     x637

10).
    23500
     x448

11).
    98594
     x150

12).
    49069
     x216

13).
    48078
     x880

14).
    63376
     x148

15).
    88418
     x623

16).
    33588
     x214

17).
    84692
     x551

18).
    96015
     x471

19).
    91917
     x566

20).
    20915
     x981

21).
    83438
     x631

22).
    92283
     x785

23).
    68664
     x719

24).
    76733
     x482

25).
    42372
     x943

26).
    64629
     x728

27).
    68820
     x814

28).
    72223
     x683

29).
    36562
     x359

30).
    45706
     x138

Multiplication 5-Digit By 3-Digit Numbers

Day: ⏱ **Time:**   **Score:**   **/30**

1). 78554
   x466
_____

2). 92886
   x754
_____

3). 74209
   x745
_____

4). 68496
   x796
_____

5). 75634
   x755
_____

6).
  62661
   x739
_____

7).
  88575
   x883
_____

8).
  41743
   x234
_____

9).
  67275
   x690
_____

10).
  51856
   x854
_____

11).
  15843
   x731
_____

12).
  67362
   x933
_____

13).
  25690
   x841
_____

14).
  53296
   x804
_____

15).
  79236
   x863
_____

16).
  34497
   x679
_____

17).
  70653
   x399
_____

18).
  51881
   x164
_____

19).
  71590
   x588
_____

20).
  55443
   x223
_____

21).
  33555
   x917
_____

22).
  34355
   x322
_____

23).
  18879
   x675
_____

24).
  26836
   x768
_____

25).
  29065
   x437
_____

26).
  66333
   x805
_____

27).
  22715
   x571
_____

28).
  76010
   x537
_____

29).
  25213
   x470
_____

30).
  30537
   x121
_____

Multiplication 5-Digit By 3-Digit Numbers

Day: 🕐 **Time:**   **Score:**   /30

1). 58914
     x239

2). 60539
     x840

3). 37810
     x119

4). 24985
     x779

5). 84490
     x417

6). 86021
     x651

7). 96778
     x382

8). 50117
     x978

9). 39425
     x717

10). 70670
     x401

11). 41932
     x387

12). 65364
     x502

13). 56711
     x992

14). 17906
     x937

15). 28590
     x580

16). 24948
     x702

17). 27322
     x475

18). 77977
     x336

19). 64488
     x219

20). 82961
     x767

21). 95009
     x848

22). 67631
     x286

23). 11322
     x725

24). 77396
     x897

25). 65191
     x444

26). 86132
     x401

27). 48148
     x800

28). 97320
     x250

29). 86567
     x934

30). 24350
     x754

Multiplication 5-Digit By 3-Digit Numbers

Day: 🕐 **Time:**   **Score:**   **/30**

1). 75194
  x864

2). 73708
  x476

3). 48808
  x933

4). 94520
  x775

5). 56244
  x605

6).
  71926
  x209

7).
  40614
  x303

8).
  79684
  x234

9).
  99584
  x621

10).
  66170
  x485

11).
  63703
  x966

12).
  84342
  x665

13).
  82059
  x677

14).
  94929
  x930

15).
  80859
  x282

16).
  39079
  x830

17).
  55987
  x107

18).
  96721
  x886

19).
  47598
  x356

20).
  33398
  x211

21).
  23362
  x326

22).
  61424
  x446

23).
  97097
  x707

24).
  20897
  x609

25).
  81243
  x432

26).
  72302
  x826

27).
  74839
  x507

28).
  95224
  x269

29).
  93527
  x676

30).
  97045
  x209

Multiplication 5-Digit By 3-Digit Numbers

Day: ⏱ **Time:**   **Score:**   **/30**

1).   21585
     x637

2).   14244
     x919

3).   72057
     x965

4).   20336
     x154

5).   29267
     x592

6).   58831
     x783

7).   88520
     x179

8).   82502
     x322

9).   23979
     x337

10).   27603
     x993

11).   28833
     x290

12).   47316
     x283

13).   44709
     x465

14).   58245
     x871

15).   40682
     x567

16).   46642
     x477

17).   69672
     x344

18).   24244
     x120

19).   81445
     x276

20).   75817
     x333

21).   96976
     x549

22).   31811
     x197

23).   98479
     x865

24).   59897
     x831

25).   93093
     x153

26).   68557
     x647

27).   35886
     x143

28).   34421
     x412

29).   69620
     x511

30).   82556
     x979

Answer Key

Page1

1570	1011	653	1008	1349
609	1539	1352	1491	1169
974	1604	662	1225	1212
971	1059	1291	1242	1179
921	1246	1626	1911	1163
1563	1557	1195	440	781

Page2

1749	756	572	1252	1007
1896	1422	1411	1027	926
1069	736	903	1202	758
1132	1249	1069	260	459
1298	851	1425	931	735
1366	1560	1385	1781	385

Page3

1458	1117	853	1342	549
825	1614	1410	1321	1014
1515	1465	588	1347	1608
1236	718	1058	1423	1288
541	918	1114	1057	654
742	1003	1092	924	923

Page4

527	752	1269	1047	958
1122	799	866	1857	873
1324	824	1335	609	941
1445	1269	648	1357	402
954	1556	1175	919	1323
939	844	1840	1381	637

Page5

1229	1038	1143	957	900
1065	1686	1265	1263	772
748	1719	1382	1575	1004
589	1585	527	1263	436
1019	1218	1196	906	1002
1113	682	962	922	699

Page6

9666	16077	12879	17220	5623
12423	9326	13642	8974	16178
12420	7558	10052	4685	12810
11121	9554	12216	9972	5723
12392	13533	17144	9194	6456
14010	13225	10680	13526	11669

Page7

16658	17281	6002	3218	8284
7098	5489	9948	13873	10181
8904	16724	18176	9125	17023
6015	8357	5301	6999	14571
9670	13302	10722	9698	8664
15236	18599	10318	18810	13859

Page8

10185	8533	13826	11649	13394
8608	5055	12433	15998	8867
11618	9227	10066	6609	10611
7645	10435	10669	9100	4394
11255	12945	5895	12473	12909
10477	13585	10401	10214	6155

Page9

11494	6283	16509	14816	13180
7963	9355	10385	10632	7715
6890	15734	2633	11518	16128
4426	14502	7167	11001	5172
9013	18930	16518	9062	12206
10528	14691	7519	10927	4375

Page10

11452	13491	16709	16991	10077
12768	14058	10804	16009	15198
13712	7845	15904	12157	10773
11867	12282	15540	8915	6967
7433	11352	8097	11894	8885
14193	6297	16753	15462	11784

Page11

27653	41056	68220	77035	38894
60823	22668	101517	81556	69222
77700	78621	64259	75398	73745
31566	38934	71529	54285	33130
95623	67087	65102	55183	36776
21084	49705	28077	80360	39009

Page12

43479	53787	104167	43184	49598
96736	73896	80347	19453	16860
87446	104773	69655	22458	82000
72025	89013	68329	50053	57136
93292	54199	38521	79988	105489
46975	103955	15811	21718	77332

Page13

23426	64464	21203	29784	57909
24653	62900	76447	54568	53942
55001	45437	18831	76529	61755
48489	15394	45843	96971	26283
32411	84662	22160	91274	99153
92179	68957	93735	102173	39984

Page14

42015	91903	82833	54548	54532
102119	94755	104655	54852	51111
95859	79045	101199	32256	89370
24860	79948	14442	24255	90607
61018	55251	90335	83580	84160
21805	91440	20061	46892	51981

Page15

47758	81614	86870	20761	65671
81738	75813	100111	104698	90921
90715	50340	107837	84043	61301
19716	26206	99518	40355	56740
96721	76260	42367	105731	20585
57739	59457	68432	71919	16403

Page16

02951	171194	69569	71268	106325
60379	95387	168309	82078	114266
09116	92291	81877	148688	131459
33542	35596	79501	161375	132750
88313	103468	115467	56035	62098
82869	87821	30043	85185	97646

Page17

161773	108193	128021	104895	87954
117096	98585	117113	98883	152405
108147	132016	160060	130479	93918
112250	140352	103397	55296	51466
152971	119324	96371	114280	152018
159881	64662	74516	107196	48350

Page18

134115	85150	58165	94250	128805
128580	151409	97336	171613	74981
45762	142824	105046	168371	78076
124177	109463	101024	80444	69044
31653	74827	107872	132511	99900
72560	139661	173161	101237	96084

71984	107166	169255	56444	105545
64730	93745	169976	86295	53414
93384	100019	99846	107483	111179
47025	168794	89379	120000	94956
139107	48436	162667	64944	144071
67478	120532	173428	138956	81737

75332	105011	151900	85745	67736
146438	103862	160898	134487	143288
149287	180107	151364	126202	36438
88136	85738	70425	127586	98006
140808	121745	76845	72880	56900
146999	85403	128491	101759	162355

20	101	432	247	161
228	514	391	24	111
113	207	242	146	44
382	644	122	18	1
147	4	92	248	99
10	270	410	74	78

380	14	363	789	366
284	239	74	310	48
351	151	76	513	237
213	6	328	72	270
3	460	152	258	361
95	139	158	51	151

25	709	32	341	652
314	172	3	37	222
99	60	93	254	11
569	509	292	531	113
29	211	93	34	9
410	334	43	265	651

69	64	419	177	180
50	90	7	104	262
4	59	88	407	235
16	213	60	120	43
494	151	535	99	159
49	7	145	357	54

232	154	29	785	10
55	14	554	176	41
5	147	47	153	17
26	174	117	123	478
229	133	53	371	48
8	610	348	267	20

2008	3035	4789	3505	5250
43	3531	404	4503	258
8614	47	2341	2078	74
3412	939	1404	251	1225
1339	226	266	230	886
827	498	1015	4444	182

1120	5945	1497	1048	1121
1322	367	631	2873	1108
4735	647	5275	3654	20
7819	1101	283	5774	8227
898	1233	1701	712	1173
691	592	2586	4547	9

556	2236	2959	3130	224
549	1398	6512	476	2790
3989	2655	3712	358	899
4894	467	1487	3266	5187
5631	387	3866	368	4268
4732	6	286	1018	1386

200	812	92	223	2330
2164	154	260	1446	5969
224	2037	95	2031	2157
1486	3649	2684	7	7266
871	6831	3026	662	1982
626	4096	2408	1771	6210

5734	202	1277	2386	3875
1829	1922	6509	1798	916
2511	1669	5975	613	195
4123	1305	6014	3886	27
210	1166	4841	840	574
2821	478	158	923	5423

12556	63504	73025	65849	83896
81991	56364	45979	30190	66409
72366	57949	23434	52800	67743
91243	9307	51697	84361	93644
38393	30364	78179	48669	61350
89745	5062	5749	4678	13033

53674	19660	13192	46374	71041
92634	25042	43744	58931	63053
41412	14517	92228	41873	75044
49264	11088	27020	21121	17848
13246	20312	54500	40965	25960
21830	44299	70260	86357	63364

25166	59742	4865	63861	71063
15870	57577	45702	74727	54009
11976	45434	27047	62217	36060
57981	9026	64412	40186	84027
43783	62394	13283	46968	15541
46149	24081	42361	51490	35630

78703	39466	13624	32257	27193
40690	31292	47481	29315	53007
45353	45920	49547	5088	68562
49335	7436	67895	29954	62143
33313	59996	61070	36260	42319
7369	20520	77183	8695	60961

49453	18511	65023	44570	90469
39719	3263	11173	66821	86533
36120	44565	75574	61924	57320
20560	60230	8346	35448	36672
91709	52882	40453	62955	56564
73262	19622	24087	24672	85318

33894	16363	20066	10637	69832
32810	42887	30116	3860	19416
29522	6312	2468	2626	63559
36581	10088	60130	10681	22035
11349	13308	82563	2058	61090
9096	1252	66444	431	39092

Page37

33373	44862	43962	13062	73232
49702	3097	23104	15661	9941
4122	12939	38086	13245	2393
80896	6271	61931	23161	1274
21217	9206	30577	4921	19311
39416	54313	31252	5316	38186

Page38

3750	15324	3619	33954	24378
3612	34995	74114	7929	1832
33362	66212	56034	28269	37292
53106	16287	29890	30760	12762
1324	55838	28860	2319	12548
31207	21344	64118	4283	2744

Page39

10261	14537	18257	21339	48389
29617	24712	57498	39840	25514
5120	17519	13165	29243	0
4336	1806	5802	372	43421
34870	79181	51959	23194	56075
30296	8138	17146	11569	12689

Page40

29334	13504	13334	67770	31872
4604	36372	4156	6387	2822
5298	31524	13862	5371	2651
36178	25979	54087	47186	10445
5964	28166	50563	2286	7138
43283	22405	1153	54719	18127

Page41

44261	37827	11052	48640	2680
43704	47536	76976	18046	29695
3569	57870	19404	36700	18760
21208	11421	63609	1188	24522
29136	24300	29724	67545	62552
9992	75344	14238	63840	25045

Page42

12402	10504	11352	26520	12222
7098	49848	11508	6910	26379
1660	71888	22080	5240	10982
23256	23625	43929	17180	78360
76768	2342	19696	25989	14296
40528	37968	10610	11890	12012

Page43

32632	10602	7472	55671	14784
2068	9522	42340	66920	60613
13852	19557	37792	27573	29500
29955	6824	17820	17907	7644
57920	54496	12520	46890	19050
10887	26469	9624	12574	29925

Page44

54714	55736	9897	10036	59481
43267	36570	25407	8778	12044
15008	60776	68744	24008	20490
1703	8856	72848	29071	24464
40600	4460	67552	23604	5944
65648	23317	7280	49420	5031

Page45

44645	11256	25467	47803	12948
37110	8939	5859	41512	12222
9080	3745	16594	8193	23639
9760	4580	56448	22338	34032
31452	41448	30198	22672	47585
35819	13164	38964	2492	35190

Page46

246588	191404	694632	77625	452900
104721	398124	531304	288996	16919
455903	333408	623063	297010	55764
231714	557244	450755	92358	76370
88799	59510	692370	310416	84008
140056	331815	258159	51996	246240

Page47

13343	41340	142128	695592	382404
545940	253833	90652	254016	90816
201516	99176	493759	157700	156802
373116	412209	464460	238098	384232
457030	332808	133956	52168	36666
197449	289785	28682	711801	65846

Page48

154455	308915	182880	92815	226240
345354	87185	251005	155844	65402
114558	173616	234180	296514	191664
301990	683207	86823	131112	246675
123438	712809	131112	208410	161214
247480	181848	131528	411859	171566

Page49

237852	315720	197204	305070	60186
156030	454086	572112	530080	136984
565821	57420	657414	268268	376110
506640	510202	512225	140754	333224
116217	572047	338485	155296	78747
734424	599344	70595	97257	101085

Page50

667773	53518	775629	80316	510713
99867	595260	305340	89728	455175
202527	62466	169582	64814	298590
274209	61512	577479	465276	426790
184197	78743	308904	355880	119936
133370	382685	78175	428455	465528

Page51

480716	153024	856622	442018	43408
516273	41256	336897	388700	319616
54304	475595	709152	123318	423544
46020	342022	715532	205989	288036
73200	311432	193344	97734	388124
99134	259686	704154	523500	273812

Page52

584948	158202	269438	417582	108255
85560	284790	677026	556764	345114
149611	282848	77310	577674	155839
224424	517270	512540	257152	399528
532196	88880	88352	444060	92304
119010	57696	418018	154704	141394

Page53

144056	528055	231912	337348	421254
339242	401219	29900	113967	492324
629280	152577	28350	343039	47250
394320	407622	97630	93648	253118
259686	762420	44080	307396	516166
548146	150252	186412	532610	323268

Page54

225984	25845	265050	211572	603477
389545	660113	285432	461871	149072
204876	321510	86922	213400	775392
143920	620720	287926	40990	286165
452664	577727	602985	169200	356021
250380	342420	384768	46002	85816

115882	244400	752851	157833	233951
145899	251988	693441	750060	820272
296471	201872	241836	494000	315836
85792	141081	195660	202500	183328
229194	563880	216972	234912	657957
615611	289440	119476	331254	538080

2906410	70750116	40048869	9997900	65182850
12974307	36050316	17754792	27471262	10528000
14789100	10598904	42308640	9379648	55084414
7187832	46665292	45223065	52025022	20517615
52649378	72442155	49369416	36985306	39956796
47049912	56019480	49328309	13125758	6307428

36606164	70036044	55285705	54522816	57103670
46306479	78211725	9767862	46419750	44285024
11581233	62848746	21605290	42849984	68380668
23423463	28190547	8508484	42094920	12363789
30769935	11062310	12743325	20610048	12701405
53398065	12970265	40817370	11850110	3694977

14080446	50852760	4499390	19463315	35232330
55999671	36969196	49014426	28267725	28338670
16227684	32812728	56257312	16777922	16582200
17513496	12977950	26200272	14122872	63631087
80567632	19342466	8208450	69424212	28944804
34538932	38518400	24330000	80853578	18359900

64967616	35085008	45537864	73253000	34027620
15032534	12306042	18646056	61841664	32092450
61537098	56087430	55553943	88283970	22802238
32435570	5990609	85694806	16944888	7046978
7616012	27395104	68647579	12726273	35096976
59721452	37943373	25615256	63224252	20282405

13749645	13090236	69535005	3131744	17326064
46064673	15845080	26565644	8080923	27409779
8361570	13390428	20789685	50731395	23066694
22248234	23967168	2909280	22478820	25247061
53239824	6266767	85184335	49774407	14243229
44356379	5131698	14181452	35575820	80822324

Page61

2085	6269R1	16442R3	89277	4397
9200R2	10369R2	17380R2	6933R3	9725R6
11127R1	35145R1	5006R3	41845R1	5972R4
12226R3	24101R2	83549	9777R2	6486
16076R2	10042R1	19300R2	26150R1	7774R8
12544R5	27430R1	12456R5	4058R4	10853
6435R4	5118R4	4566R1	3046R5	1306R3
1532R3	24764R1	4710	8229R3	10594R5

Page62

10418R2	41007	3593R5	12192R4	5926R1
20511R2	20127	7250R4	27137R1	12451R2
31783	33368R1	76857	9731R3	8128
17648R1	14081	5419R1	1516R4	7789R3
5092R1	7708R6	4735	59118	12725R1
15775R3	10441	13284R1	6831R3	17122R3
12968R6	7843R1	39531	4631R7	6150R3
13157R2	15942	9217R2	8998R5	2028R7

Page63

9285R3	6747R3	9761	14960	62158
10912R4	8883R2	9007R5	13643R2	24622
11616	10630R8	21369	10464R3	10000R2
34043	4239R4	5174R3	3754R4	5450R5
21011	47231R1	1470R1	2049R1	8748R3
12970R4	15350R1	9428R7	6256R2	15962R1
13972R1	19801R1	9322R3	11873	21459R2
9424R2	25556R2	7874R5	17390R1	81584

Page64

11995R6	9818R6	1712R1	9170R1	12399R1
12235R4	5962	25335	12985R3	8189R5
8567R3	2483R1	6727R6	3243R2	43155R1
17870R1	18462	19165	7776R1	31593
80666	11533R5	7360	12269R3	18089R2
8011R4	5303R2	93521	12031R1	15455R1
13229	8847	14633	3135	9814R7
6122	11905	8106R5	4066R2	15461R3

Page65

12859R1	2871	3088R5	1349R5	23423
20714	14191R1	4682R3	6513R3	13911R3
20544R2	19805R3	2100R1	12686R2	4070R2
16510R2	6140R2	3082R4	28831	20743R2
2717R4	18763R1	5604R2	5919R1	9457R4
2602R4	22136	4036R3	1994R3	56718
12981R2	37812	89824	17218	13491R3
8749R1	9673R3	12958	18362R3	13283R2

Page66

116785R2	106991R1	960694	53423R3	116966
136242	199717R1	162963R2	74172R2	276312
99426R6	132142R3	50211R3	902630	218001
77226R1	526046	35032R6	212711R2	42833R6
129458R3	900555	34068R4	141983	292233
99923R3	35376R2	519459	142764	108560R4
422655	93400R3	139818	72435R4	95982R3
244171R3	45653R3	404889	56243R1	103794R4

Page67

318617	225788R2	94952R1	60338R1	108826R2
1514R4	252297	34462R4	30660R3	175474R1
80345R2	105011R1	144511R2	48986R3	155839R1
48326R1	54905R1	143645R1	284720R1	329333
00915R5	142951R2	277655	103019R1	43960
7722R2	132636R1	25478R6	134895R4	159158R1
3968R1	91502R5	267566	74487R1	59754R1
3920R3	105622R2	450408R1	124915R2	225643R2

Page68

124102R2	20468R5	64546	131764	465592R1
38473R2	134505R2	97382R1	65974R2	133083R3
126153	107868	444911R1	81670R3	48836R1
33736R4	96651R2	119633R1	148980R2	174867R4
44094R1	131474R1	113690R1	108423R4	49084R4
694093	22072R7	494957R1	148072R2	70428R2
467699	118735R4	72033R1	132248	39751R1
301955R2	103942R1	223283R1	352908	73070R2

Page69

132578R2	56177R2	131092R1	30422R4	170636
83806R4	64999R7	42217R1	164341R3	25476R5
99367R1	65890R3	210319	84538	90232R4
95215R4	68804R2	69855R2	71798R2	63551R4
88084R1	27334	85324R2	54544R5	462235
195648R1	162120R1	383634R1	24450R4	37304R7
96876R2	218160	86454R4	114519R2	128893R1
91110R1	151056R5	129677R3	201320R2	763836

Page70

3142R2	195153	164828	18133R1	477862
7744R1	23276R4	499891	36282R3	156726R2
527612	211894	182546	82136R2	202535
47251R1	331813	110889R5	393363	256399R1
30540R1	188998R3	43165R1	72074R1	154504
216312	756091	460581R1	95618	311044R1
1744R5	22930R1	31022R4	260386	59404R5
234880	318240	177080R3	200504R1	146875R3

Page71

83R32	313R15	207R20	284R2	122R8
199R20	16R59	24R88	92R40	86R68
195R41	100R51	71R82	62R48	349R10
45R17	67R48	182R4	57R70	68R7
242R14	404R15	292R18	91R63	43R21
116R12	112R43	80R29	153R23	59R76
402R12	63R90	109R3	170R21	107R33
87R25	140R26	109R6	452R13	70R31

Page72

147R35	48R41	98R46	56R47	16R23
261R14	81R72	63R82	91R32	212R6
48R9	19R37	51R32	90R28	101R32
61R4	49R50	144R7	54R29	67R34
362R8	79R34	149R40	159R8	613R2
37R8	127R54	160R48	98R11	20R79
13R57	92R33	32R74	173R21	351R20
90R15	67R7	119R5	79R8	180R18

Page73

186R18	428R15	86R62	414R5	114R73
169R17	171R26	64R66	34R80	134R48
558R10	52R8	130R54	13R27	59R16
94R21	86R6	86R23	64R10	154R31
29R63	200R9	31R28	30R26	99R29
158R4	42R38	216R27	54R16	128R21
54R50	291R1	521R12	117R38	407R17
131R5	98R44	25R31	146R25	55R55

Page74

92R23	99R44	45R12	197R5	178R20
61R51	82R6	105R41	40	420R9
73R27	95R12	218R7	150R12	348R5
97R19	67R73	66R7	98R14	37R36
58R40	51R11	53R18	204R3	114R28
434R3	221R1	130	134R56	74R45
239R30	35R41	38R1	36R45	60R98
115R1	35R41	164R2	74R3	23R9

Page75

153R5	66R29	64R7	46R40	122R5
107R20	416R14	79R28	269R17	91R28
35R31	79R29	110R24	140R4	90R56
24R56	46R3	149R16	290R11	78R5
224R3	40R31	41R57	99R57	208R19
294R18	153R6	112R58	118R3	271R1
217R4	80R16	355R7	107R4	32R22
169R47	55R14	609R2	55R55	120R20

Page76

1072R76	706R35	5037R15	2587R12	1304R24
1327R57	499R5	868R19	1632R19	160R66
895R19	366R18	307R7	432R11	2924R18
644R46	925R8	1126	876R13	475R1
608R10	612R71	1694R5	477R1	572R28
2492R28	975R12	1227R3	668R13	611R20
425R14	4693R3	222R20	1292R12	1113R13
1691R31	2354R30	2404R1	1032R29	6638R8

Page77

1044R71	1535R17	506R6	1592R33	373R3
805R73	6645R4	1094R2	1058R9	1083R11
122R72	5543R7	288R30	3308R14	1161R42
1520R46	421R46	1318R2	1100R21	184R47
3617R5	567R28	995R12	594R22	4495R3
723R30	631R30	1047R53	724R27	943R52
2451R12	1211R27	1767R17	7883	841R7
6736R3	1490R6	735R64	1870R27	306R11

Page78

465R29	895R59	298R12	692R22	484R1
2622R27	7232R7	954R10	4332R12	291R3
780R71	1842R38	1347R14	808R2	720R3
537R7	903R4	189R57	139R29	1176R
501R77	675R19	2679R23	2122	167R3
752R1	1446R14	313R49	653R7	1251R
1580R2	1129R78	1605R14	800R20	1826R
505R12	1055R41	1132R19	767R62	605R8

Page79

1213R16	360R45	1218R19	580R11	2827R12
935R14	1027R37	3257R6	1594R14	1665R15
3332R4	2395R1	483R54	1045R20	968R56
1719R18	447R8	1697R18	1178R14	1096R64
1776R7	2562R14	341R45	572R70	1253R29
301R15	165R10	1253R13	1898R38	1369R23
848R70	1494R29	1367R10	393R63	572R49
2296R5	242R45	112R82	239R67	1886R8

Page80

2080R20	712R59	248R12	1413R52	1474R36
2088R12	976R11	2227R10	799R69	117R54
355R59	1821R11	228R36	2105R2	611R47
1283R18	1466R32	821R4	1349R19	2364R7
915R52	736R35	8640R5	427R49	2131R13
395R83	2020R15	282R44	1142R35	515R2
1147R45	1226R13	1650R34	251R33	3450R8
270R64	3857R13	866R22	3046R7	174R20

MULTIPLICATION TABLE

1 × 1 = 1	1 × 2 = 2	1 × 3 = 3	1 × 4 = 4
2 × 1 = 2	2 × 2 = 4	2 × 3 = 6	2 × 4 = 8
3 × 1 = 3	3 × 2 = 6	3 × 3 = 9	3 × 4 = 12
4 × 1 = 4	4 × 2 = 8	4 × 3 = 12	4 × 4 = 16
5 × 1 = 5	5 × 2 = 10	5 × 3 = 15	5 × 4 = 20
6 × 1 = 6	6 × 2 = 12	6 × 3 = 18	6 × 4 = 24
7 × 1 = 7	7 × 2 = 14	7 × 3 = 21	7 × 4 = 28
8 × 1 = 8	8 × 2 = 16	8 × 3 = 24	8 × 4 = 32
9 × 1 = 9	9 × 2 = 18	9 × 3 = 27	9 × 4 = 36
10 × 1 = 10	10 × 2 = 20	10 × 3 = 30	10 × 4 = 40
11 × 1 = 11	11 × 2 = 22	11 × 3 = 33	11 × 4 = 44
12 × 1 = 12	12 × 2 = 24	12 × 3 = 36	12 × 4 = 46

1 × 5 = 5	1 × 6 = 6	1 × 7 = 7	1 × 8 = 8
2 × 5 = 10	2 × 6 = 12	2 × 7 = 14	2 × 8 = 16
3 × 5 = 15	3 × 6 = 18	3 × 7 = 21	3 × 8 = 24
4 × 5 = 20	4 × 6 = 24	4 × 7 = 28	4 × 8 = 32
5 × 5 = 25	5 × 6 = 30	5 × 7 = 35	5 × 8 = 40
6 × 5 = 30	6 × 6 = 36	6 × 7 = 42	6 × 8 = 48
7 × 5 = 35	7 × 6 = 42	7 × 7 = 49	7 × 8 = 56
8 × 5 = 40	8 × 6 = 48	8 × 7 = 56	8 × 8 = 64
9 × 5 = 45	9 × 6 = 54	9 × 7 = 63	9 × 8 = 72
10 × 5 = 50	10 × 6 = 60	10 × 7 = 70	10 × 8 = 80
11 × 5 = 55	11 × 6 = 66	11 × 7 = 77	11 × 8 = 88
12 × 5 = 60	12 × 6 = 72	12 × 7 = 84	12 × 8 = 96

1 × 9 = 9	1 × 10 = 10	1 × 11 = 11	1 × 12 = 12
2 × 9 = 18	2 × 10 = 20	2 × 11 = 22	2 × 12 = 24
3 × 9 = 27	3 × 10 = 30	3 × 11 = 33	3 × 12 = 36
4 × 9 = 36	4 × 10 = 40	4 × 11 = 44	4 × 12 = 48
5 × 9 = 45	5 × 10 = 50	5 × 11 = 55	5 × 12 = 60
6 × 9 = 54	6 × 10 = 60	6 × 11 = 66	6 × 12 = 72
7 × 9 = 63	7 × 10 = 70	7 × 11 = 77	7 × 12 = 84
8 × 9 = 72	8 × 10 = 80	8 × 11 = 88	8 × 12 = 96
9 × 9 = 81	9 × 10 = 90	9 × 11 = 99	9 × 12 = 108
10 × 9 = 90	10 × 10 = 100	10 × 11 = 110	10 × 12 = 120
11 × 9 = 99	11 × 10 = 110	11 × 11 = 121	11 × 12 = 132
12 × 9 = 108	12 × 10 = 120	12 × 11 = 132	12 × 12 = 144

The End

As you turn the final page of this book, remember that addition, subtraction, multiplication, and division are not just operations on numbers - they are powerful tools for problem-solving, reasoning, and critical thinking. Whether you're a student, a teacher, or simply someone who loves learning, let this book be a starting point for your journey of mathematical discovery.

Take what you've learned here and apply it to real-world situations. Share your knowledge with others, and never stop exploring the beauty and wonder of mathematics. Remember, every problem you solve and every concept you understand brings you one step closer to realizing your full potential. So go forth, be curious, be brave, and most importantly, keep learning!

The journey of growth and learning is ongoing, and there will always be new challenges and obstacles to overcome. But it's important to remember that every setback is an opportunity to learn and grow stronger.

So I encourage you to keep pushing forward, even when the road ahead seems difficult. Surround yourself with positive influences and never lose sight of your vision for the future.

Thank you for joining me on this journey, and I wish you all the best in your own personal growth and success.

Sincerely,

Thank you for your purchase, if you loved that book or not, Don't hesitate to give your opinion (constructive ;-)) and your ideas for improvement after your purchase, because I really want to offer quality, it takes five seconds and helps small businesses like ours

William. Education

Made in the USA
Las Vegas, NV
12 April 2024